Copper in Drinking Water

COMMITTEE ON COPPER IN DRINKING WATER

BOARD ON ENVIRONMENTAL STUDIES AND TOXICOLOGY

COMMISSION ON LIFE SCIENCES

NATIONAL RESEARCH COUNCIL

NATIONAL ACADEMY PRESS
WASHINGTON, D.C.

NATIONAL ACADEMY PRESS 2101 Constitution Ave., N.W. Washington, D.C. 20418

NOTICE: The project that is the subject of this report was approved by the Governing Board of the National Research Council, whose members are drawn from the councils of the National Academy of Sciences, the National Academy of Engineering, and the Institute of Medicine. The members of the committee responsible for the report were chosen for their special competences and with regard for appropriate balance.

This project was supported by Cooperative Agreement No. CR827277-01 between the National Academy of Sciences and the U.S. Environmental Protection Agency. Any opinions, findings, conclusions, or recommendations expressed in this publication are those of the author(s) and do not necessarily reflect the view of the organizations or agencies that provided support for this project.

Library of Congress Card Catalog Number 00-102668
International Standard Book Number 0-309-06939-4

Additional copies of this report are available from:

National Academy Press
2101 Constitution Ave., NW
Box 285
Washington, DC 20055

800-624-6242
202-334-3313 (in the Washington metropolitan area)
http://www.nap.edu

Copyright 2000 by the National Academy of Sciences. All rights reserved.

Printed in the United States of America

THE NATIONAL ACADEMIES

National Academy of Sciences
National Academy of Engineering
Institute of Medicine
National Research Council

The **National Academy of Sciences** is a private, nonprofit, self-perpetuating society of distinguished scholars engaged in scientific and engineering research, dedicated to the furtherance of science and technology and to their use for the general welfare. Upon the authority of the charter granted to it by the Congress in 1863, the Academy has a mandate that requires it to advise the federal government on scientific and technical matters. Dr. Bruce M. Alberts is president of the National Academy of Sciences.

The **National Academy of Engineering** was established in 1964, under the charter of the National Academy of Sciences, as a parallel organization of outstanding engineers. It is autonomous in its administration and in the selection of its members, sharing with the National Academy of Sciences the responsibility for advising the federal government. The National Academy of Engineering also sponsors engineering programs aimed at meeting national needs, encourages education and research, and recognizes the superior achievements of engineers. Dr. William A. Wulf is president of the National Academy of Engineering.

The **Institute of Medicine** was established in 1970 by the National Academy of Sciences to secure the services of eminent members of appropriate professions in the examination of policy matters pertaining to the health of the public. The Institute acts under the responsibility given to the National Academy of Sciences by its congressional charter to be an adviser to the federal government and, upon its own initiative, to identify issues of medical care, research, and education. Dr. Kenneth I. Shine is president of the Institute of Medicine.

The **National Research Council** was organized by the National Academy of Sciences in 1916 to associate the broad community of science and technology with the Academy's purposes of furthering knowledge and advising the federal government. Functioning in accordance with general policies determined by the Academy, the Council has become the principal operating agency of both the National Academy of Sciences and the National Academy of Engineering in providing services to the government, the public, and the scientific and engineering communities. The Council is administered jointly by both Academies and the Institute of Medicine. Dr. Bruce M. Alberts and Dr. William A. Wulf are chairman and vice chairman, respectively, of the National Research Council.

COMMITTEE ON COPPER IN DRINKING WATER

RICHARD BULL (*Chair*), Battelle Pacific Northwest Division, Richland, Wash.
MICHAEL ASCHNER, Wake Forest University, Winston-Salem, N.C.
GEORGE BREWER, University of Michigan, Ann Arbor, Mich.
EDWARD HARRIS, Texas A&M University, College Station, Tex.
CARL KEEN, University of California, Davis, Calif.
KARL KELSEY, Harvard University, Boston, Mass.
F. WILLIAM SUNDERMAN, JR., Middlebury College, Middlebury, Vt.
JOYCE TSUJI, Exponent, Bellevue, Wash.
LAUREN ZEISE, California Environmental Protection Agency, Oakland, Calif.

Staff

CAROL A. MACZKA, Project Director
MICHELLE C. CATLIN, Postdoctoral Research Associate
RUTH E. CROSSGROVE, Editor
JUDITH L. ESTEP, Senior Program Assistant
LAURA T. HOLLIDAY, Senior Program Assistant
MIRSADA KARALIC-LONCAREVIC, Information Specialist

Sponsor: U.S. Environmental Protection Agency

BOARD ON ENVIRONMENTAL STUDIES AND TOXICOLOGY

GORDON ORIANS (*Chair*), University of Washington, Seattle, Wash.
DONALD MATTISON (*Vice Chair*), March of Dimes, White Plains, N.Y.
DAVID ALLEN, University of Texas, Austin, Tex.
INGRID C. BURKE, Colorado State University, Fort Collins, Colo.
WILLIAM L. CHAMEIDES, Georgia Institute of Technology, Atlanta, Ga.
JOHN DOULL, The University of Kansas Medical Center, Kansas City, Kan.
CHRISTOPHER B. FIELD, Carnegie Institute of Washington, Stanford, Calif.
JOHN GERHART, University of California, Berkeley, Calif.
J. PAUL GILMAN, Celera Genomics, Rockville, Md.
BRUCE D. HAMMOCK, University of California, Davis, Calif.
MARK HARWELL, University of Miami, Miami, Fla.
ROGENE HENDERSON, Lovelace Respiratory Research Institute, Albuquerque, N.M.
CAROL HENRY, Chemical Manufacturers Association, Arlington, Va.
BARBARA HULKA, University of North Carolina, Chapel Hill, N.C.
JAMES F. KITCHELL, University of Wisconsin, Madison, Wisc.
DANIEL KREWSKI, University of Ottawa, Ottawa, Ont.
JAMES A. MACMAHON, Utah State University, Logan, Utah
MARIO J. MOLINA, Massachusetts Institute of Technology, Cambridge, Mass.
CHARLES O'MELIA, Johns Hopkins University, Baltimore, Md.
WILLEM F. PASSCHIER, Health Council of the Netherlands
KIRK SMITH, University of California, Berkeley, Calif.
MARGARET STRAND, Oppenheimer Wolff Donnelly & Bayh, LLP, Washington, D.C.
TERRY F. YOSIE, Chemical Manufacturers Association, Arlington, Va.

Senior Staff

JAMES J. REISA, Director
DAVID J. POLICANSKY, Associate Director and Senior Program Director for Applied Ecology
CAROL A. MACZKA, Senior Program Director for Toxicology and Risk Assessment
RAYMOND A. WASSEL, Senior Program Director for Environmental Sciences and Engineering
KULBIR BAKSHI, Program Director for the Committee on Toxicology
LEE R. PAULSON, Program Director for Resource Management
ROBERTA M. WEDGE, Program Director for Risk Analysis

COMMISSION ON LIFE SCIENCES

MICHAEL T. CLEGG *(Chair)*, University of California, Riverside, Calif.
PAUL BERG *(Vice Chair)*, Stanford University, Stanford, Calif.
FREDERICK R. ANDERSON, Cadwalader, Wickersham & Taft, Washington, D.C.
JOANNA BURGER, Rutgers University, Piscataway, N.J.
JAMES E. CLEAVER, University of California, San Francisco, Calif.
DAVID EISENBERG, University of California, Los Angeles, Calif.
JOHN EMMERSON, Fishers, Ind.
NEAL FIRST, University of Wisconsin, Madison, Wisc.
DAVID J. GALAS, Keck Graduate Institute of Applied Life Science, Claremont, Calif.
DAVID V. GOEDDEL, Tularik, Inc., South San Francisco, Calif.
ARTURO GOMEZ-POMPA, University of California, Riverside, Calif.
COREY S. GOODMAN, University of California, Berkeley, Calif.
JON W. GORDON, Mount Sinai School of Medicine, New York, N.Y.
DAVID G. HOEL, Medical University of South Carolina, Charleston, S.C.
BARBARA S. HULKA, University of North Carolina, Chapel Hill, N.C.
CYNTHIA KENYON, University of California, San Francisco, Calif.
BRUCE R. LEVIN, Emory University, Atlanta, Ga.
DAVID LIVINGSTON, Dana-Farber Cancer Institute, Boston, Mass.
DONALD R. MATTISON, March of Dimes, White Plains, N.Y.
ELLIOT M. MEYEROWITZ, California Institute of Technology, Pasadena, Calif.
ROBERT T. PAINE, University of Washington, Seattle, Wash.
RONALD R. SEDEROFF, North Carolina State University, Raleigh, N.C.
ROBERT R. SOKAL, State University of New York, Stony Brook, N.Y.
CHARLES F. STEVENS, The Salk Institute, La Jolla, Calif.
SHIRLEY M. TILGHMAN, Princeton University, Princeton, N.J.
RAYMOND L. WHITE, University of Utah, Salt Lake City, Utah

Staff

WARREN R. MUIR, Executive Director
JACQUELINE K. PRINCE, Financial Officer
BARBARA B. SMITH, Administrative Associate
LAURA T. HOLLIDAY, Senior Program Assistant

OTHER REPORTS OF THE
BOARD ON ENVIRONMENTAL STUDIES AND TOXICOLOGY

Ecological Indicators for the Nation (2000)
Waste Incineration and Public Health (1999)
Hormonally Active Agents in the Environment (1999)
Research Priorities for Airborne Particulate Matter: I. Immediate Priorities and a Long-Range Research Portfolio (1998); II. Evaluating Research Progress and Updating the Portfolio (1999)
Ozone-Forming Potential of Reformulated Gasoline (1999)
Risk-Based Waste Classification in California (1999)
Arsenic in Drinking Water (1999)
Brucellosis in the Greater Yellowstone Area (1998)
The National Research Council's Committee on Toxicology: The First 50 Years (1997)
Toxicologic Assessment of the Army's Zinc Cadmium Sulfide Dispersion Tests (1997)
Carcinogens and Anticarcinogens in the Human Diet (1996)
Upstream: Salmon and Society in the Pacific Northwest (1996)
Science and the Endangered Species Act (1995)
Wetlands: Characteristics and Boundaries (1995)
Biologic Markers (5 reports, 1989-1995)
Review of EPA's Environmental Monitoring and Assessment Program (3 reports, 1994-1995)
Science and Judgment in Risk Assessment (1994)
Ranking Hazardous Waste Sites for Remedial Action (1994)
Pesticides in the Diets of Infants and Children (1993)
Issues in Risk Assessment (1993)
Setting Priorities for Land Conservation (1993)
Protecting Visibility in National Parks and Wilderness Areas (1993)
Dolphins and the Tuna Industry (1992)
Hazardous Materials on the Public Lands (1992)
Science and the National Parks (1992)
Animals as Sentinels of Environmental Health Hazards (1991)
Assessment of the U.S. Outer Continental Shelf Environmental Studies Program, Volumes I-IV (1991-1993)
Human Exposure Assessment for Airborne Pollutants (1991)
Monitoring Human Tissues for Toxic Substances (1991)
Rethinking the Ozone Problem in Urban and Regional Air Pollution (1991)
Decline of the Sea Turtles (1990)

Copies of these reports may be ordered from
the National Academy Press
(800) 624-6242
(202) 334-3313
www.nap.edu

Preface

IN 1991, the U.S. Environmental Protection Agency (EPA) promulgated a maximum contaminant level goal (MCLG) of 1.3 mg/L for copper in drinking water. Some states and municipalities have difficulty maintaining copper levels below the MCLG primarily because of the characteristics of their water. On the basis of recent data from epidemiological studies, questions have been raised about the validity of the science on which the MCLG is based and whether that level is appropriate. In response to those questions, the U.S. Congress requested that the National Research Council (NRC) independently evaluate the toxicological, epidemiological, and exposure data on copper and determine whether EPA's MCLG is scientifically valid.

In this report, the Committee on Copper in Drinking Water of the NRC reviews the validity of the scientific basis for EPA's MCLG. The committee reviewed the available toxicological, epidemiological, and exposure data (from food and water) and evaluated the appropriateness of the critical study, end points of toxicity, and uncertainty factors used by EPA in the derivation of the MCLG for copper. The committee was also asked to identify data gaps and make recommendations for future research.

This report has been reviewed in draft form by individuals chosen for their diverse perspectives and technical expertise in accordance with procedures approved by the NRC's Report Review Committee for reviewing NRC and Institute of Medicine reports. The purpose of this independent review is to provide candid and critical comments that will assist the NRC in making the published report as sound as possible and to ensure that the report meets institutional standards for objectivity, evidence, and responsiveness to the study charge. The review comments and draft manuscripts remain confidential to protect the integrity of the deliberative process. The committee wishes to thank the following individuals, who are neither officials nor employees of the NRC, for their participation in the review of this report: John L. Emmerson, Eli Lilly (retired) Portland, Oregon; Diane W. Cox, University of Alberta; Henry A. Anderson, Wisconsin Division of Public Health; Dennis Thiele, University of Michigan; Richard Stevens, University of Connecticut; George Becking, Phoenix OHC; George Cherian, University of Western Ontario; Vincent Piccirillo, NPC, Incorporated; Joseph Rodricks, The Life Sciences Consultancy.

The individuals listed above have provided many constructive comments and suggestions. It must be emphasized, however, that responsibility for the final content of this report rests entirely with the authoring committee and the NRC.

The committee gratefully acknowledges the following individuals for providing background information and for making presentations to the committee: Edward Ohanian, Joyce Donohue, and James Taft, all of EPA; Debbie Reed, of the office of Senator J. Robert Kerrey (Nebraska); and Allison Yates and Paula Trumbo, of the NRC Food and Nutrition Board. The committee also acknowledges Scott Baker, of the International Copper Association; Ken Poirier, of Toxicology Excellence for Risk Assessment; and Magdalena Araya, of the University of Chile, for information on the international tri-site acute copper study.

We are grateful for the assistance of the NRC staff in preparing the report. Staff members who contributed to this effort are Carol A. Maczka, senior program director for toxicology and risk assessment; Michelle Catlin, postdoctoral research associate; Ruth E. Crossgrove, editor; Judy Estep, senior project assistant; Laura Holliday, senior project assistant; and Mirsada Karalic-Loncarevic, information specialist.

Finally, I would like to thank all the members of the committee for their dedicated efforts throughout the development of this report.

                                        Richard J. Bull, Ph.D.
                                        Chairman
                                        Committee on Copper in Drinking Water

Contents

EXECUTIVE SUMMARY	1
1 INTRODUCTION	9
Chemical and Physical Properties, *11*	
Sources of Copper in Drinking Water, *12*	
Committee's Approach to its Charge, *12*	
Structure of the Report, *13*	
References, *14*	
2 PHYSIOLOGICAL ROLE OF COPPER	16
Essentiality, *16*	
Biochemistry and Physiology, *17*	
Factors Affecting Bioavailability, *21*	
Conclusions, *26*	
Recommendations, *26*	
References, *27*	
3 HEALTH EFFECTS OF COPPER DEFICIENCIES	33
Teratogenesis of Copper Deficiency, *33*	
Health Effects of Copper Deficiencies in Adults, *42*	
Conclusions, *43*	
References, *45*	

4 Disorders of Copper Homeostasis — 51

Menkes Disease, *51*
Occipital Horn Syndrome, *53*
Wilson Disease, *55*
Genetic Characteristics of Wilson and Menkes Diseases, *56*
Heterozygotes for Wilson Disease, *58*
Aceruloplasminemia, *62*
Tyrolean Infantile Cirrhosis, *64*
Indian Childhood Cirrhosis, *64*
Idiopathic Copper Toxicosis, *65*
Other Genetic Disorders, *65*
Disease-Induced Changes in Copper Homeostasis, *66*
Conclusions, *67*
Recommendations, *68*
References, *69*

5 Health Effects of Excess Copper — 78

Acute Toxicity, *78*
Chronic Toxicity, *87*
Animal Studies, *95*
Conclusions, *111*
Recommendations, *112*
References, *113*

6 Risk Characterization — 127

Copper Deficiency, *127*
Copper Toxicity from Single or Short-Term Exposure, *130*
Copper Toxicity from Chronic Exposure, *132*
Chronic Copper Exposure Through Tap Water, *138*
Dietary Contribution and Total Copper Intake, *139*
Conclusions, *141*
Recommendations, *144*
References, *145*

Copper in Drinking Water

Executive Summary

THE U.S. Environmental Protection Agency (EPA) is required under the Safe Drinking Water Act (SDWA) to establish the concentrations of contaminants permitted in public drinking-water supplies. Enforceable standards are to be set at concentrations that have no observable adverse health effects with adequate margins of safety and are attainable with the use of the best available technology. The maximum contaminant level (MCL) is the term used for enforceable standards. The maximum contaminant level goal (MCLG) is based on available scientific data. It is not a regulatory requirement, and, in principle, it might not be attainable with current technology. According to the SDWA, the MCL should be set as close to the MCLG as is feasible with available technology. The MCLG for copper will be used by EPA as a basis for establishing the MCL.

Copper is a naturally occurring element that is present in drinking water. Stagnation of water in pipes and plumbing fixtures containing copper and copper alloys in distribution systems and household plumbing allows leaching and increases water copper levels. Characteristics of the water, including increased acidity, increased temperature, and reduced hardness, can increase the leaching of copper into the water.

Acute ingestion of excess copper in drinking water is associated with adverse health effects, including acute gastrointestinal disturbances, and chronic ingestion of copper can lead to liver toxicity in sensitive populations. The current EPA MCLG of 1.3 milligrams per liter (mg/L) for copper in drinking water is based on the need to protect against adverse gastrointestinal effects. Some states and municipalities have difficulty maintaining copper levels below the MCLG primarily because of the characteristics of their water. Questions have been raised about the validity of the science on which the MCLG is based and whether that level is appropriate. Some believe that the level might be unnecessarily low, and others believe that some individuals might have adverse health effects with copper levels at or below the current MCLG.

THE CHARGE TO THE COMMITTEE

At the direction of Congress, EPA asked the National Research Council (NRC) to review independently the scientific and technical basis for EPA's MCLG of 1.3 mg/L for copper in drinking water. For the review, the NRC convened the Committee on Copper in Drinking Water, whose members have expertise in the fields of toxicology, epidemiology, pathology, pharmacology, genetics, physiology, medicine, public health, exposure assessment, nutrition, chemistry, biostatistics, and risk assessment. The committee reviewed available toxicological, epidemiological, and exposure data (from food and water) and determined the appropriateness of the critical study used for deriving the MCLG, end points of toxicity, and uncertainty factors used by EPA in the derivation of the MCLG for copper. The committee was also asked to identify data gaps and make recommendations for future research. The committee was not asked to address risk-management issues.

THE COMMITTEE'S APPROACH TO ITS CHARGE

The committee evaluated data relating to key elements of the risk-assessment process that led to the current MCLG. The key elements are hazard identification, dose response, exposure assessment, and risk characterization. The current MCLG is based on gastrointestinal effects following acute exposure to copper. However, effects on the liver have been observed with chronic exposure in sensitive populations. Therefore, in this report, the committee reviewed information on the health effects of copper exposure in humans following both acute and chronic oral exposure. The committee also evaluated data on the mechanisms of action of copper toxicity, the health effects associated with copper deficiencies, and factors affecting the bioavailability of copper—all data that could affect the risk assessment. The toxicity of copper was used as the basis for evaluating the safety of copper concentrations in drinking water, but the essential need for copper as a micronutrient (i.e., the dietary essentiality of copper) was taken into account when considering uncertainty factors.

To provide background information on relevant issues in copper toxicity, various government representatives and trade organizations gave presentations to the committee. Those presentations included representatives of EPA, who sponsored the report, the office of Senator J. Robert Kerrey (Nebraska), and the International Copper Association. The committee also heard from scientists with relevant expertise in copper toxicity and from the Institute of Medicine's Food and Nutrition Board regarding the essentiality of copper.

The committee recognized that exposure to copper can occur from multiple sources; however, emphasis was placed on the human health effects associated with exposure through ingestion of drinking water. The committee did consider the contribution of copper in food. Exposure via inhalation or dermal routes (e.g., shower or bath) were not addressed.

THE COMMITTEE'S EVALUATION

Health Effects of Excess Copper

The primary health effects following acute exposure to copper are gastrointestinal disturbances, including nausea and vomiting. Direct irritation of the stomach by copper ions appears to underlie the acute toxic response. Although a number of cases of copper toxicity in humans have been reported in the literature, few are suitable for identifying minimal concentrations at which effects are seen. Those studies provide qualitative information on gastrointestinal end points but are not suitable for establishing an MCLG. Currently, the MCLG is based on one such case report. A recently published controlled study in humans, however, was designed to determine the copper concentration in water at which gastrointestinal disturbances occur. That study indicates that gastrointestinal symptoms arise from drinking water with copper at approximately 3 mg/L. Information provided to the committee by the International Copper Association on a controlled study of gastrointestinal effects of copper appears consistent with the data in the published study.

The main concern regarding chronic exposure to excess copper is liver toxicity, especially in sensitive populations. A number of chronic cases of liver toxicity have been reported. A model for those chronic effects might be derived from patients with Wilson disease, which causes abnormal copper regulatory mechanisms that result in accumulation of excess copper. The liver and brain are targets of copper toxicity in patients with Wilson disease. Although the mechanisms by which chronic copper exposure causes damage are not fully understood, evidence suggests that as copper regulatory mechanisms are surpassed, large amounts of copper are released into the bloodstream, and oxidative damage results. Excess dietary copper has not been associated with adverse effects on reproduction and development in humans. The only reported association between dietary copper and cancer is in patients with Wilson disease, in which the association appears to be secondary to liver toxicity.

In general, animal models provide qualitative insight into the toxicology of copper, but they are of limited value for establishing dose-response relationships in humans. However, animal strains that are sensitive to copper because of genetic alterations provide valuable information on the effects and mechanisms of copper-induced toxicity in genetically sensitive populations.

Recommendations

The committee recommends that studies be conducted to establish the frequency and characteristics of gene defects in humans with Wilson disease. Characterization of the genetic basis of infant and childhood copper toxicosis should also be undertaken. Genetic animal models should be used to investigate the physiological role of copper and as models of genetically sensitive human populations. Such research should be designed to determine the association between liver toxicity and copper in sensitive populations, including Wilson-disease individuals, heterozygotes for the Wilson-disease gene, and other populations (e.g., Tyrolean infantile cirrhosis (TIC), Indian copper toxicosis (ICT), and idiopathic copper cirrhosis (ICC)) (see Sensitive Populations section below). Epidemiological studies of populations who have been chronically exposed to elevated copper should also be carried out to determine the nature and frequency of chronic effects, especially in sensitive populations.

Physiological Role of Copper

Copper is an essential micronutrient that has multiple metabolic functions. Severe copper deficiencies are associated with cardiac, bone, immune, and central-nervous-system problems. Ingested copper is readily absorbed from the intestines and transported to the liver. Specific transport molecules for copper control its movement from the liver to other tissues, and enzymes regulate its excretion into the bile. A number of factors can affect the bioavailability of copper. The percentage of copper that is absorbed changes with the amount of copper needed by the body, with age, and with dietary factors. Dietary factors include the amount of amino acids in the diet, competing metal ions, and the amount of plant versus animal protein that an individual ingests. The mechanisms by which copper is absorbed and distributed to the tissues in the body and by which copper concentrations are regulated are not fully understood.

Recommendations

The committee recommends that further research be carried out to determine the mechanisms that underlie the absorption, distribution, and regulation of copper in the body. Continued research on the role of copper in the body is warranted, with an emphasis on the consequences of both copper deficiency and copper toxicity. Such research could provide mechanistic data and information on the interactions between copper and other factors that could be used to refine risk assessments.

Sensitive Populations

The committee concludes that there are disorders in copper homeostasis in some individuals that are important to consider in the regulation of copper in drinking water. Those include Wilson disease, heterozygote carriers of the gene for Wilson disease, Tyrolean infantile cirrhosis (TIC), Indian childhood cirrhosis (ICC), and idiopathic copper toxicosis (ICT).

Wilson disease is a genetic disorder that has a recessive inheritance pattern—that is, two mutated copies of the gene must be present to manifest the full disease. The frequency of Wilson disease in the United States is relatively low (1 in 40,000).

Individuals who have only one mutated copy of the gene for Wilson disease (i.e., heterozygote carriers) also have abnormal copper regulation and can have a build-up of copper in the body. It can be calculated from the frequency of the disease that 1% of the general population is heterozygotic for a Wilson gene defect. Unlike Wilson-disease patients, heterozygous individuals are largely unrecognized in the population. The committee concludes that those individuals, perhaps identified by examining families with a medical history of Wilson disease, should be considered in establishing copper drinking-water standards.

TIC, ICC and ICT are all syndromes in which infants or young children are afflicted with liver disease. All three syndromes appear to have a genetic component. There is evidence in cases of TIC and ICC and in some cases of ICT that an increased ingestion of copper precipitated the disease. Those cases represent other groups who should be considered when regulating copper in drinking water.

Recommendations

Populations that are sensitive to excess copper in drinking water should be investigated, especially those with genetic abnormalities that might increase the risk of developing copper toxicity. Novel mechanisms that might be important in copper toxicity and imbalance should be examined.

Risk Characterization

In characterizing the risks of copper, the committee noted that adverse health effects are associated with both copper deficiency and copper excess. The committee considered the dietary essentiality of copper when discussing the uncertainty factors that are appropriate in setting an MCLG. Because the MCLG is intended to be protective against copper toxicity, the committee concluded that the MCLG for drinking water should be based on toxicity, not copper deficiency.

The main toxic end points of concern to the committee were nausea and vomiting following acute exposure to copper in drinking water, and liver effects in sensitive populations following chronic exposure. For acute effects, the committee concluded that experimental studies in humans are the most appropriate for deriving the MCLG. Although the precise concentrations at which acute effects occur are difficult to determine, the one relevant published study showed that gastrointestinal effects occurred following acute exposure at and above 3 mg/L. In establishing the MCLG for gastrointestinal effects from acute exposure, the fact that the effects are mild and not life-threatening, and that the data are from humans are important considerations.

In sensitive populations, liver toxicity can occur following chronic exposure to excess copper. The available data are plagued by imprecise exposure measurements, but there is some indication that sensitive infants might be at risk for liver toxicity at copper concentrations of approximately 3 mg/L of drinking water.

Survey data indicate that numerous water systems have first-draw-water copper concentrations in selected homes above the current MCLG. Some copper concentrations are above 3 mg/L, indicating the potential for copper toxicity. Because of the low probability that a sensitive individual would consume a sufficient volume of first-draw water at a high copper concentration, toxicity is unlikely to occur often. For toxicity to occur, first, a sensitive individual must be in a household with high concentrations of copper in water, and second, that individual must be the one who consumes the first-draw water.

Recommendations

The committee recommends that the MCLG for copper in drinking water be based on the toxic effects of copper, rather than on copper deficiency. Issues that should be considered in establishing adjustment and uncertainty factors for acute effects are that copper is an essential micronutrient, that the GI effects are not severe or life-threatening, that the effect level is based on human studies and case reports, and that the effect level appears to be at the lower part of the dose-response curve, where the majority of the population is nonresponsive.

Given the potential risk for liver toxicity in individuals with polymorphisms in genes involved in copper homeostasis, the committee recommends that the MCLG for copper not be increased at this time.

Additional information on total copper doses received from drinking water is needed before systemic chronic toxicity can be evaluated in susceptible populations. Better quantification of the frequency and characterization of copper-sensitive populations should be undertaken. When the above information on sensitive populations is obtained, the MCLG for copper should be re-evaluated.

1

Introduction

UNDER the Safe Drinking Water Act, the U.S. Environmental Protection Agency (EPA) is required to establish the concentrations of contaminants that are permitted in public drinking-water supplies. Specifically, Section 1412 of the act, as amended in 1986, requires EPA to publish maximum-contaminant-level goals (MCLGs) and promulgate national primary drinking-water regulations (e.g., MCLs) for contaminants in drinking water that might cause any adverse effect on human health and that are known or expected to occur in public water systems. MCLGs are not regulatory requirements but are health goals set at concentrations at which no known or expected adverse health effects occur and the margins of safety are adequate. The MCLG for copper will be used by EPA as a basis for establishing the MCL. MCLs are enforceable standards that are to be set as close as possible to the MCLG with the use of the best technology available.

Copper is an essential micronutrient (Underwood 1977; Goyer 1991). The Food and Nutrition Board (FNB) recommends dietary copper intake for adults of 1.5-3.0 milligrams (mg) per day (NRC 1989). The Institute of Medicine's (IOM) FNB is reviewing the recommendations. Acute ingestion of excess copper in drinking water can cause gastrointestinal (GI) tract disturbances and chronic ingestion can lead to liver toxicity in sensitive populations. In 1991, EPA promulgated an MCLG of 1.3 mg per liter (L) for copper in drinking water to protect against adverse GI tract effects. That value is based on a case study (Wyllie 1957) of nurses who consumed an alcoholic beverage that was contaminated with copper. In the study, a dose of 5.3 mg was found to cause GI symptoms. Based on an intake of 2 L per day, a concentration of 2.65 mg/L was determined to be the minimal dose at which symptoms could occur. That value was "divided by a safety factor of 2 in recognition of its essentiality" to yield the copper MCLG of 1.3 mg/L (Donohue 1997).

Several U.S. states (such as Nebraska and Delaware) have measured copper concentrations in drinking water that exceed the MCLG for copper because of the leaching of copper from plumbing. On the basis of recent data from epidemiological studies which show no adverse effects at higher levels, questions have been raised about the validity of the science on which the MCLG is based, and whether that level is appropriate. While some have argued that the level might be too conservative, others have argued that some individuals might experience adverse effects with copper levels at or below the current MCLG (Sidhu et al. 1995). A provisional drinking-water guideline of 2 mg/L was proposed for copper by the World Health Organization (WHO 1993). The basis for that value is not clear; however, an interpretation of the derivation is provided by Galal-Gorchev and Herrman (1996). The drinking-water guideline appears to originate from the provisional maximum tolerable daily intake (PMTDI) value established by the Joint FAO/WHO Expert Committee on Food Additives (JECFA). On the basis of the lack of adverse effects or copper accumulation in normal individuals with typical dietary copper intakes and the "considerable margin" between normal intakes and those with adverse effects, JECFA established a PMTDI of 0.05 mg of copper per kilogram (kg) of body weight per day (WHO 1967). Subsequent re-evaluation of that intake dose did not provide any basis for JECFA to change the recommendation. When determining the provisional guidelines for copper in drinking water, WHO assumed that a 60-kg adult would drink 2 L of water per day and that 10% of the PMTDI would come from drinking water. WHO then established the provisional drinking-water guideline of 1.5 mg/L (WHO 1991), which was later rounded to 2 mg/L (WHO 1993). Several states have also recommended exposure limits different from those proposed by EPA (summarized by the ATSDR) (ATSDR 1990). Olivares and Uauy (1996) and Fitzgerald (1998) discuss the drinking water standards for copper established by different agencies and governments.

In response to concern regarding the scientific validity of EPA's MCLG, the U.S. Congress requested that the administrator of EPA enter into a contract with the NRC to conduct a comprehensive study of the effects of copper in drinking water on human health. In response to that request, the NRC convened the Committee on Copper in Drinking Water. The committee's expertise is in the fields of toxicology, epidemiology, pathology, pharmacology, genetics, physiology, medicine, public health, exposure assessment, nutrition, chemistry, biostatistics, and risk assessment. The committee was charged to review independently the appropriateness of the

EPA MCLG of 1.3 mg/L for copper. The specific tasks of the committee were to (1) evaluate toxicology, epidemiology, and exposure data (from food and water); and (2) determine whether the critical study, end point of toxicity, and uncertainty factors used by EPA in the derivation of the MCLG for copper are appropriate. The committee was also asked to identify data gaps and make recommendations for future research. The committee was not asked to address, nor did it address, risk-management issues, and the committee did not attempt to derive an MCL for copper.

CHEMICAL AND PHYSICAL PROPERTIES

Copper is number 29 in the Periodic Table of Elements. Copper has a ground state electronic configuration of $3d^{10}4s^1$ and occurs in the environment in three major valence states: copper metal (Cu^o), Cu(I) and Cu(II). As a member of the $3d$ transition metal series, copper and six other metals in the series—chromium, iron, cobalt, manganese, nickel, and zinc—constitute the bulk of essential metals in biological systems. Its transition metal properties are caused by partially filled $3d$ orbitals, a characteristic of all metals (with the exception of zinc) in the series. The partially filled ($3d^9$) orbital permits Cu(II) complexes to be highly colorful. The gemstone turquoise is copper aluminum hydroxyphosphate in which Cu^{2+} ions provide the blue-green color. Metals in the series also have defined spatial geometries. Cu(I) usually exists in a tetrahedral arrangement, whereas Cu(II) complexes most often are square planar. Loss of the single $4s$ electron gives rise to Cu(I), a weak oxidant with a closed $3d^{10}$ shell and a featureless absorption spectrum. Cu(II) is the more stable form. Copper is commonly found in ores. The principal ore minerals are chalcopyrite ($CuFeS_2$), cuprite (Cu_2O), and malachite $Cu_2(CO_3)(OH)_2$.

The southwestern United States is one of the world's largest producers of copper. Because minerals such as malachite are plentiful and cheap, metallic copper has been used in many industrial applications, ranging from coins and ornamental jewelry to relatively inexpensive plumbing fixtures. The metal has the properties of malleability, ductility, and electrical conductivity, which make it a preferred choice in the building industry for hot- and cold-water pipes, electrical wires, hose nozzles, and castings. Brass, an alloy of copper and zinc, has been used in cooking ware and musical instruments. Bronze, an alloy of copper with about 5-10% tin, is used in castings and marine equipment. Metallic copper is basically unstable and is subject to corrosion and leaching. It is a mistake, therefore, to consider copper metal or any of its alloys as impervious to environmental conditions. Oxygen in the air reacts slowly with brass, forming the familiar green coating on fixtures. Although copper metal tends not to leach in neutral solutions, organic solvents, and detergents, acid solution effectively leaches traces of the metal as cupric ions. Thus, valves in soft-drink dispensing machines are subject to the corrosive effects of carbonic acid and have been shown to be a source of toxic cupric ions. Cupric ions are much less soluble in alkaline solution because of the formation of highly insoluble cupric hydroxide, $Cu(OH)_2$. Amine compounds and ammonia, however, form complex tetracupric amines (e.g., $[Cu(NH_3)_4]^{2+}$), which are highly soluble in both acids and bases.

SOURCES OF COPPER IN DRINKING WATER

Copper is a natural element with widespread distribution. It is present in the environment in different valence states and in different complexes. The form of copper affects its solubility; therefore, the copper forms present in water will be different from those found in food. In rivers, copper is generally adsorbed to insoluble particles or complexed with inorganic ligands (Florence et al. 1992). In drinking water, copper is generally free in solution.

Human activities can release copper into the environment, especially to the land. Mining operations, along with incineration, are the main sources of copper release. Release into water occurs from weathering of soil, industrial discharge, sewage-treatment plants, and antifouling paints (IPCS 1998). The concentrations of copper in drinking water can be greatly increased during the distribution of drinking water. Many pipes and plumbing fixtures contain copper, which can leach into the drinking water. Characteristics of the water that can increase the leaching of copper include low pH, high temperature, and reduced hardness. Electrolysis of copper from pipes can result from using household pipes to ground appliances. The length of time that the water has been sitting stagnant in the pipes can also greatly increase the concentration of copper to several milligrams per liter in the water (EPA 1994). The concentration of copper is much higher in first-draw water than in water after the tap has been flushed.

COMMITTEE'S APPROACH TO ITS CHARGE

The committee evaluated data related to hazard identification, dose response, and risk characterization-key elements of the risk-assessment process-that address the protective nature of the current MCLG. Specifically, the committee reviewed information on the health effects of excess copper exposure in humans following acute and chronic oral exposure. The current MCLG is based on GI effects following acute exposure rather than chronic exposure. However, chronic-exposure effects in the liver have been observed in sensitive populations. Therefore, in this report, the committee addresses the effects of acute and chronic exposure to copper. The committee also evaluated information that could affect the risk assessment. That included data on the mechanism of action of copper toxicity, the health effects associated with copper deficiencies, and factors affecting the bioavailability of copper. The toxicity of copper was used as the basis for evaluting a "safe" level of copper in drinking water, but the essentiality of copper was taken into account when considering uncertainty factors.

To gather background information on relevant issues in copper toxicity, various government representatives and trade organizations gave presentations to the committee. Those presentations included representatives of the EPA, who sponsored the report, the office of Senator J. Robert Kerrey of Nebraska, and the International Copper Association. The committee also heard from scientists with relevant expertise in copper toxicity and from the IOM's Food and Nutrition Board regarding the essentiality of copper.

The committee recognized that exposure to copper can occur from multiple sources and considered the role of food in copper intake. Exposure via inhalation and dermal routes, although possible from copper in drinking water were not addressed.

STRUCTURE OF THE REPORT

The remainder of this report is organized in five chapters. Chapter 2 discusses the physiological role of copper, including its essentiality, biochemistry, and physiology. Factors affecting the bioavailability of copper are also identified. In Chapter 3, the health effects associated with copper deficiencies are presented. Disorders of copper homeostasis are described in Chapter 4. Chapter 5 addresses the health effects following acute and chronic exposure to excess copper. Particular attention is given to the health effects in sensitive populations, including individuals who are heterozygous for genetic disorders of copper homeostasis. In addition, Chapter 5 presents toxicity data from experimental animals and discusses the appropriateness of animal models for studying the underlying mechanism and toxicity of copper in humans. Chapter 6 characterizes the risks associated with acute and chronic exposure to excess copper and provides a discussion on the appropriate use of uncertainty factors and the public-health implications of the narrow margin of safety of the MCLG for copper.

REFERENCES

ATSDR (Agency for Toxic Substances and Disease Registry). 1990. Toxicological Profile for Copper. U.S. Department of Health and Human Services, Public Health Service, ATSDR, Atlanta, GA.

Donohue, J. 1997. New Ideas after Five Years of the Lead and Copper Rule: A Fresh Look at the MCLG for Copper. Pp. 265-272 in Advances in Risk Assessment of Copper in the Environment, G.E. Lagos and R. Badilla-Ohlbaum, eds. Santiago, Chile: Catholic University of Chile.

EPA (U.S. Environmental Protection Agency). 1994. Drinking Water; Maximum Contaminant Level Goal and National Primary Drinking Water Regulation for Lead and Copper. Fed. Regist. 59(125):33860-33864.

Fitzgerald, D.J. 1998. Safety guidelines for copper in water. Am. J. Clin. Nutr. 67(5 Suppl.):1098S-1102S.

Florence, T.M., G.M. Morrison and J.L. Stauber. 1992. Determination of trace element speciation and the role of speciation in aquatic toxicity. Sci. Total. Environ. 125:1-13.

Galal-Gorchev, H. and Herrman, J.L. 1996. Letter to A.C. Kolbye, Jr., editor of Regulatory and Pharmacology, on the evaluation of copper by the Joint FAO/WHO Expert Committee on Food Additives from WHO, Sept.12, 1996.

Goyer, R.A. 1991. Toxic effects of metals. Pp. 623-680 in Casarrett and Doull's Toxicology: The Basic Science of Poisons, 4th Ed., M.O. Amdur, J. Doull, and C.D. Klaassen, eds. New York: Pergamon Press.

IPCS (International Programme on Chemical Safety). 1998. Copper. Environmental Health Criteria 200. Geneva, Switzerland: World Health Organization.

NRC (National Research Council). 1989. Recommended Dietary Allowances, 10th Ed. Washington, D.C.: National Academy Press.

Olivares, M. and R. Uauy. 1996. Limits of metabolic tolerance to copper and biological basis for present recommendations and regulations. Am. J. Clin. Nutr. 63(5 Suppl.):846S-52S.

Sidhu, K.S., D.F. Nash, and D.E. McBride. 1995. Need to revise the national drinking water regulation for copper. Regul. Toxicol. Pharmacol. 22:95-100.

Underwood, E.J. 1977. Copper. Pp.56-108 in Trace Elements in Human and Animal Nutrition, 4th Ed. New York: Academic Press.

WHO (World Health Organization). 1967. Specifications for the Identity and Purity of Food Additives and Their Toxicological Evaluation: Some Emulsifiers and Stabilizers and Certain Other Substances. Tenth report of the Joint FAO/WHO Expert Committee on Food Additives. FAO Nutrition Meetings Series, No. 43, WHO Technical Report Series No. 373. Geneva, Switzerland: World Health Organization.

WHO (World Health Organization). 1991. Revision of the WHO Guidelines for drinking-water quality. Report of the Second Review Group Meeting on Inorganics. Brussels, Belgium, Oct. 14-18, 1991. Document number WHO/PEP/91.32. Geneva, Switzerland: World Health Organization.

WHO (World Health Organization). 1993. Guidelines for Drinking-Water Quality. Vol.1. Recommendations, 2nd Ed. Geneva, Switzerland: World Health Organization.

Wyllie, J. 1957. Copper poisoning at a cocktail party. Am. J. Publ. Health 47:617.

2

Physiological Role of Copper

THIS chapter begins by discussing the basis of the essentiality of copper. The kinetics of copper in the body and select roles of copper at the cellular and molecular level are described. In addition, factors influencing the bioavailability of copper are presented.

ESSENTIALITY

Copper is both essential and toxic to living systems. As an essential metal, copper is required for adequate growth, cardiovascular integrity, lung elasticity, neovascularization, neuroendocrine function, and iron metabolism. An average adult human ingests about 1 mg of copper per day in the diet; about half of which is absorbed (Harris 1997). An expert committee of the World Health Organization recommends 30 micrograms (µg) per kilogram (kg) of body weight per day, which equates to about 2 mg per day for an average adult (WHO 1996). The Food and Nutrition Board (FNB) recommends dietary copper intake of 1.5-3.0 mg per day for adults (NRC 1989). Copper is obligatory for enzymes involved in aerobic metabolism, such as cytochrome *c* oxidase in the mitochondria, lysyl oxidase in connective tissue, dopamine monooxygenase in brain, and ceruloplasmin. As a cofactor for apo-copper-zinc superoxide dismutase (apoCuZnSOD), copper protects against free-radical damage to proteins, membrane lipids, and nucleic acids in a wide range of cells and organs. Severe copper deficiencies, either gene defects due to mutations or low dietary copper intakes, although relatively rare in humans, have been linked to mental retardation, anemia, hypothermia, neutropenia, diarrhea, cardiac hypertrophy, bone fragility, impaired immune function, weak connective tissue, impaired central-nervous-system (CNS) functions, peripheral neuropathy, and loss of skin, fur (in animals), or hair color (Linder and Goode 1991; Uauy et al. 1998; Cordano 1998; Percival 1998).

BIOCHEMISTRY AND PHYSIOLOGY

Copper taken in through the diet might be absorbed partially in the stomach, where the highly acidic environment frees the bound copper ions from partially digested food particles. However, the largest portion of ingested copper passes into the duodenum and ileum, which are the major sites of absorption. As a result of complexing with amino acids, organic acids, or other chelators, a high fraction of copper is soluble in the intestinal tract. Studies on isolated segments of the duodenum suggest that copper ions enter into the mucosal cells lining the intestine by simple diffusion and exit at the basolateral surface by a different mode of transport (Bremner 1980). Recent reports indicate that there is a divalent transporter that might transport copper (Rolfs and Hedinger 1999).

Basolateral transport is markedly reduced in Menkes disease, which results in systemic copper deficiency. Studies of this disease led to the prediction of a copper-transporting adenosine triphosphatase (ATPase) in the basolateral membrane of mucosal cells. The copper-transporting ATPase presumably discharges the copper into the serosal capillaries where the copper binds to albumin and amino acids for transport to the liver via the portal circulation. From the liver, copper is transported to extrahepatic tissues by albumin, amino acids, and, to a lesser extent, ceruloplasmin (Dunn et al. 1991).

A large fraction of circulating copper is returned to the liver as ceruloplasmin-bound copper. Ceruloplasmin, a sialoglycoprotein, is constantly being secreted from the liver into the blood. When ceruloplasmin returns to the liver, the sialic acid can be removed by the outer endothelial cells followed by an endocytosis of the desialated protein via the asialoglycoprotein receptor in the liver parenchyma (Irie and Tavassoli 1986). Likewise, removal of copper from ceruloplasmin hastens its uptake by liver parenchymal cells (Holtzman and Gaumnitz 1970).

As discussed in Chapter 4, Wilson disease, a genetic disease characterized by accumulation of copper mainly in the liver and brain, attests to a potential role for ceruloplasmin biosynthesis in liver homeostasis of copper. Copper-containing fragments of ceruloplasmin are found in the bile of normal subjects and are generally absent from the bile of Wilson patients.

Although many Wilson patients do not synthesize sufficient amounts of ceruloplasmin, this decrease in biliary ceruloplasmin is observed even in Wilson patients who have reasonably normal blood concentrations of the protein. That observation suggests that biliary secretion of copper bound to ceruloplasmin accounts for most of the copper that is excreted (Iyengar et al. 1988). In aceruloplasminemia, a genetic defect in ceruloplasmin biosynthesis, ceruloplasmin in the circulation is totally absent (Harris et al. 1995). Yet, the aceruloplasminemic patient does not develop Wilson disease, nor does copper accumulate in the liver in aceruloplasminemia. Thus, factors besides ceruloplasmin are required for biliary copper excretion and for maintaining normal liver copper homeostasis in the liver. A clue to the identity of such factors has come from studies of lipofuschin-like granules that accumulate in the liver of patients who suffer from primary biliary cirrhosis and are unable to excrete copper. These granules apparently arise from primary lysosomes and might contain degradation fragments of metallothionein-bound copper (Humbert et al. 1982). Studies in Wistar rats suggest that biliary copper excretion occurs by a glutathione-dependent (rapid) phase and a glutathione-independent (slow) phase (Houwen et al. 1990). No identification has been made of which phase is associated with ceruloplasmin.

A second copper-transporting ATPase enzyme is required to transport the copper ions into the Golgi compartment for incorporation into apoceruloplasmin (Murata et al. 1995). Several types of cells have been shown to have receptors for ceruloplasmin. However, the above observation supports, but does not define a role for ceruloplasmin in copper delivery to tissues. The receptors have been found on many cell types. K562 cells are capable of engaging ceruloplasmin in vitro and transporting ceruloplasmin-bound copper to cellular enzymes (Percival and Harris 1991). Unlike transferrin, which delivers iron by a receptor-mediated endocytosis of the whole protein, ceruloplasmin protein is not endocytosed (except in hepatic cells) and delivers its bound copper to components at the cell surface. The role of the protein, therefore, stops at the cell membrane, and transport of copper to the interior requires transport proteins in the membrane. Ascorbate facilitates the release of copper, and chelators of copper prevent its absorption by the cell (Percival and Harris 1989, 1990).

When excess copper is fed to rats, it selectively accumulates in periportal and midzones of the liver lobules (Fuentealba et al. 1989), suggesting that portal blood flow determines copper disposition (Haywood 1981). In some patients with chronic cirrhosis, copper also accumulates in lysosomes (Humbert et al. 1982), suggesting that those organelles take part in copper storage or, more likely, in copper excretion when copper concentrations are high (Gross et al. 1989).[1] Metallothionein is a small cysteine-rich protein that tightly binds copper. This protein is thought to play roles in both copper storage and internal copper movement (Cousins 1985). At high concentrations, copper can stimulate metallothionein synthesis. Under normal conditions, copper exists at extremely low concentrations as the free ion in the cytosol. Copper-binding ligands, therefore, are key regulators in copper movement and adaptation to toxic effects. The ligands protect against toxicity and can help target the proteins for copper incorporation. Copper-binding ligands include glutathione (GSH), amino acids, ATP, and the recently identified copper metallochaperones. All of these ligands have been shown to transport copper within the cell from one location to another, and make copper available to intracellular enzymes (Sternlieb 1980; DiSilvestro and Cousins 1983; Holt et al. 1986; Harris 1995).

Numerous biochemical and nutritional studies have focused on the mechanisms of copper absorption and metabolism. The challenge is to explain how extremely small quantities of copper are able to pass through cell membranes and into enzymes that require copper for function. Some insight has been obtained from studies in yeast. Yeast mediate copper transport with separate low- and high-affinity systems for copper uptake (Eide 1998). The yeast gene *CTR1* (copper transport 1) was the first copper transport gene to be discovered (Dancis et al. 1994a,b). *CTR1* encodes a transmembrane protein that selectively transports Cu(I). A homologous human gene, *hCTR1*, that encodes a transporter protein was identified in HeLa cells (Zhou and Gitschier 1997). Because the above transporters recognize only Cu(I), a reductase enzyme must reduce the Cu^{2+} before transport into the cell. Two reductase genes that can reduce Cu^{2+} for transport were identified in yeast (Hassett and Kosman 1995; Georgatsou et al. 1997). A reductase system that uses NADH as an electron donor to reduce copper has been found in rat liver cells (van den Berg and McArdle 1994).

Cells in a defined culture medium take up copper. High-affinity membrane permeases allow copper ions at micromolar concentrations to move through the phospholipid bilayer and enter the cytoplasm. The rapid up-

---

[1] It is important to distinguish between free copper ions and copper as a complex with amino acid or other biocompounds. Solubility considerations suggest that copper as the free metal exists at extremely low concentrations in the cell cytosol. That makes diffusion-driven mechanisms unworkable and suggests that transfer between components must involve copper bound as a ligand to peptides or other agents that heighten solubility and reactivity (Harris 1995).

take is not ATP dependent, suggesting a passive carrier system (Schmitt et al. 1983; Tong and McArdle 1995). In general, chemical form, valence state, and relative concentrations of competing metals influence the quantity of copper that is absorbed. Select amino acids such as histidine and glutamine (Harris and Sass-Kortsak 1967), and ions, such as sodium, chloride, and bicarbonate (Alda and Garay 1990), stimulate transmembrane copper movement in vitro, whereas zinc and copper chelators, such as phytate, reduce absorption.

Glutathione is a ubiquitous cysteine-containing tripeptide, is present in millimolar quantities in liver, brain, kidney, and other tissues (Deneke and Fanburg 1989). Glutathione avidly binds copper. There is a rapid turnover of a GSH-Cu(I) complex in hepatoma cells. That turnover is consistent with GSH involvement in the reduction of Cu(II) to Cu(I), potentially facilitating its binding to metallothionein (Freedman and Peisach 1989a; Freedman and Peisach 1989b).

Mammalian and yeast cells have small polypeptides that bind Cu(I) and transfer it to selected recipients. These proteins, called copper chaperones or metallochaperones, transiently bind Cu(I) at a cysteine-rich region in the peptide chain (Valentine and Gralla 1997). Chaperones move copper ions from one location in the cell to another, sometimes crossing through organelle membranes. The chaperone promotes a rapid exchange of copper with the target proteins (Portnoy et al. 1999). Targets include cytochrome c oxidase in the mitochondria, ATP7B in the trans-Golgi, and apo-CuZnSOD in the cell cytosol. Three chaperones first described in yeast are known to have human counterparts. ATOX1 (formerly HAH1) is the human homolog of ATX1 and is abundantly expressed in all tissues. ATX1 and ATOX1 specifically deliver copper to the secretory pathway where the targets are membrane-bound copper ATPases that regulate the flow of copper into cell compartments. Examples include Ccc2p, which delivers copper to a multicopper oxidase (Fet3p) required for iron uptake in yeast, and ATP7B (the Wilson disease gene product), which delivers copper to apo-ceruloplasmin in the trans-Golgi of liver. Proper ATP7B function is essential for the excretion of copper in the bile. COX17 is a chaperone that transports copper to cytochrome oxidase in the mitochondria of yeast cells (Amaravadi et al. 1997). The human homolog is hCOX17. LYS7, a 27-kilodalton (kDa) protein that delivers copper to the apoCuZnSOD (Culotta et al. 1997, 1999), has a human counterpart designated CCS (copper chaperone for SOD). Yeast mutants that are defective in Lys7 are unable to incorporate copper into SOD1 and hence are defective for SOD1 activity. CCS is comparable in size and has a 28% sequence identity with LYS7. Both proteins contain a single MHCXXC consensus copper-binding sequence.

FACTORS AFFECTING BIOAVAILABILITY

The amount of copper which is absorbed from the diet can vary considerably depending on other dietary constituents. However, in general approximately half the copper consumed in the diet is typically absorbed by the gastrointestinal (GI) tract. Approximately two-thirds of the copper that is absorbed is rapidly secreted into the bile (Lönnerdal 1998; Turnlund 1998; Wapnir 1998) (Figure 2-1). Thus, approximately 80-90% of dietary copper is typically excreted in the feces. Thus, copper homeostasis is primarily regulated at the GI level, through biliary excretion with the kidney excreting only small amounts of copper. Small amounts of copper are also excreted in hair and sweat. The bioavailability, or the fraction of copper absorbed from the GI tract, has been shown to be influenced by the age of the individual, the amount of copper in the GI tract, and various dietary components. With respect to dietary components, copper in meat has been reported to be more bioavailable than that in vegetables. The bioavailability of copper is also expected to depend on the form of copper present (Baker et al. 1991). Absorption is higher for soluble or ionic forms than for less soluble or insoluble mineral forms or copper deposited in soil.

In adults, absorption varies according to the amount of copper in the diet. Animal studies indicate that absorption rates can be as low as about 10% with very high copper intakes, and as high as around 70% with low copper intakes. The average for typical diets in animals and humans is 30-40% (Lönnerdal 1996, 1998; Turnlund et al. 1998; Wapnir 1998). Turnlund et al. (1998) measured the amount of copper excreted by humans in the feces for 12 days after an oral dose, or intravenous dose, of copper for four subjects and five subjects, respectively. The intravenous dose allows for an estimation of the endogenous excretion into the gut. Retention from oral intake as estimated from copper in the feces was shown to be highest when dietary copper concentrations were lowest: 67% at 0.38 mg per day, 54% at 0.66 mg per day, and 44% at 2.49 mg per day (Turnlund et al. 1998). However, the estimated total percentage that was actually absorbed by the GI tract before endogenous (or biliary) excretion back to the GI tract was 77%, 73%, and 66%, respectively, for the three doses. Thus, changes in endogenous excretion (in the bile), rather than GI absorption, is the primary mechanism of action in regulating total body copper. Specifically, the change in endogenous excretion between the lowest dose and the highest dose varied between 12% and 34% (Turnlund et al. 1998), although those data might not adequately reflect long-term homeostasis because of the short-term nature of the study. Thus, increases in copper excretion in the feces with increasing copper dose is a function of decreased absorption and increased biliary secretion, the latter having a greater effect.

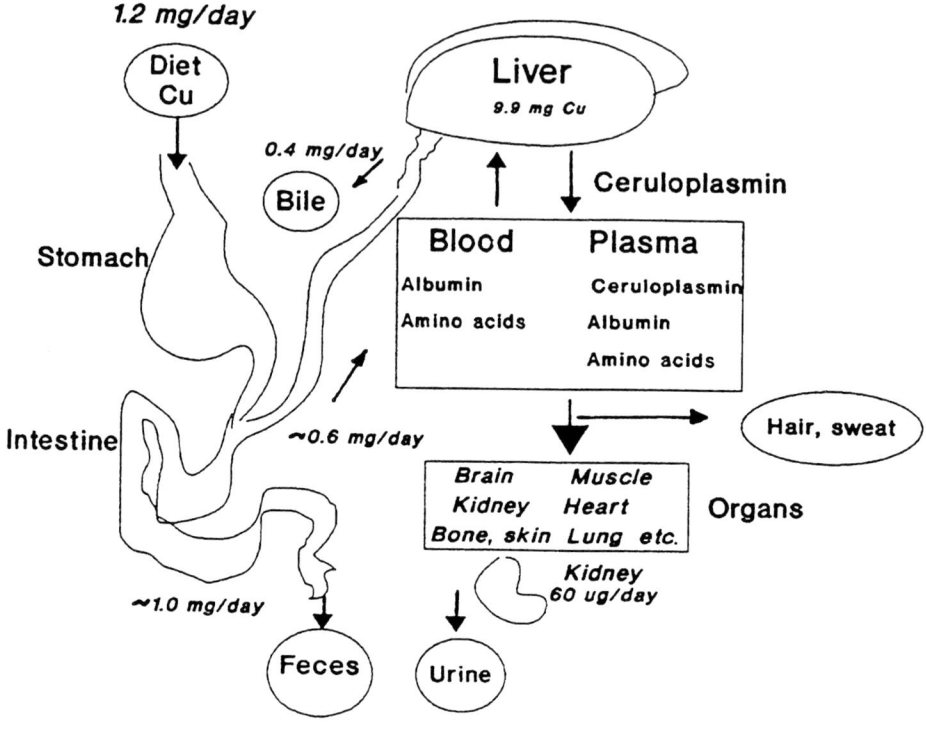

FIGURE 2-1 Overview of copper absorption, transport, and excretion. The liver receives copper from the intestine via the portal circulation and redistributes the copper to the tissue via ceruloplasmin, albumin, and amino acids. Nearly half of the copper consumed is not absorbed and passes into the feces. Another two-thirds of the daily intake is returned to the liver and released into the bile. Therefore, fecal excretion accounts quantitatively for nearly all of the copper consumed as the system endeavors to stay in balance. A small amount is excreted by the kidney via the urine, and still lesser amounts appear in hair and sweat. This interplay among the various systems maintains homeostasis and balance throughout the organism. The values in the figure are based on a dietary input of 1.2 mg per day. Source: Harris 1997. Reprinted from *Handbook of Nutritionally Essential Mineral Elements* by courtesy of Marcel Dekkar.

Effect of Age

Studies in rats indicate that the absorption and retention of copper can be particularly high in the neonatal period, and that it decreases by the time of weaning (summarized by Lönnerdal 1996, 1998). Data on absorption in human infants are sparse; however, studies on copper balance (i.e., the amount of copper in the body) show similar decreases in retention (a function of absorption and biliary excretion) of copper with age in full-term infants. Negative copper balance (a loss of copper from the body which exceeds dietary intake) can arise in preterm infants as a consequence of a reduced prenatal storage of copper and low dietary copper intake (Widdowson 1974; Sann et al. 1980; Hillman et al. 1981; Lönnerdal 1996, 1998; Cordano 1998). Both animal and human infant studies indicate that, within the intake range examined, copper absorption and retention increase linearly with intake amount, although the percentage of the dietary copper retained decreases (Lönnerdal et al. 1985; Lönnerdal 1996). For example, in 14-day-old suckling rat pups, the percentage of copper retained from a meal (0.5 mL) containing 0.2 mg of copper per liter was approximately 30%, and it was closer to 20% when the meal contained 2 mg of copper per liter. However, it is important to note that while the percentage of copper retained from the high-copper meal was lower than that from the low-copper meal, the total amount of copper retained from the high-copper meal was 7 times higher than that from the low-copper meal (Lönnerdal et al. 1985).

Due to tissue growth, and an increased expression of some copper proteins during the postnatal period, the percentage of absorbed copper retained by the body is higher in infants than in adults. Little information is available on dietary copper absorption, or copper retention, in toddlers and young children.

Dietary and Other Interactions

Early research with whole animals showed that the rate and amount of copper ions transferred across intestinal epithelia were influenced positively by dietary amino acids, but negatively by ascorbate and competing metal ions (Bremner 1980; Hogan and Rauser 1981; L'Abbe and Fischer 1984; Oestreicher and Cousins 1985; Fields et al. 1986). Chloride ions and bicarbonate appeared to stimulate cellular absorption (Alda and Garay 1990). The transport across the membrane was considered a property of transport proteins, themselves subject to antagonistic effects of competing metals, principally zinc and ferrous iron. Studies in yeast and bacteria have led to the discovery of membrane proteins that transport copper ions across cell membranes.

Copper absorption is reported to be greater in infants who ingest human milk rather than cows milk (Lönnerdal 1996 1998). That increased bioavailability might be attributed to the greater association of copper with lipids and whey in human milk, than in cow's milk, where much of the copper is bound to casein (Wapnir 1998).

The literature indicates that copper absorption is greater when diets are animal protein rather than plant protein (i.e., vegetarian) (Srikumar et al. 1992; Lönnerdal 1996; Wapnir 1998). Studies in experimental animals found that phytates and dietary fiber generally reduce copper bioavailability; however, effects on bioavailability are less clear in humans (Turnlund 1988; Lönnerdal 1996). Declines in serum copper give some indication that phytates or α-cellulose added to the diet might alter copper utilization or distribution (Turnlund 1988). In general, the effect of phytates and dietary fiber on absorption of copper appears to be less than the effect on absorption of other divalent cations, such as zinc (Wapnir 1998).

Dietary differences have been found in patients with Wilson disease. Observations of two Wilson patients on lactovegetarian diets suggest copper bioavailability is reduced by about 25% (Brewer et al. 1993b). The first of two patients on that diet was asymptomatic (i.e., normal liver function and normal slit lamp examination for Kayser-Fleischer rings) for 12 years despite having a typical average daily copper intake and no anticopper therapy. The second patient stopped using the anticopper therapy for 2 years and then switched to a lactovegetarian diet. After switching, her serum transaminase and transpeptidase activity (alanine aminotransaminase, asparate aminotransaminase, and liver γ-glutamyltranferase), which were previously elevated, showed improvement over the succeeding year. When last observed about 2 years after starting the diet, the patient remained clinically well. Other Wilson-disease patients who discontinued their anticopper therapy had serious difficulty after 3 to 4 months and serious degeneration in their condition after 1.5 years on average (Brewer and Yuzbasiyan-Gurkan 1992).

Carbohydrates, such as fructose, have been studied for their effect on copper absorption and metabolism. Fructose, or the fructose moiety of sucrose, fed to rats increased fecal and urinary excretion of copper (Lönnerdal 1996). The influence of fructose on copper balance in humans has not been well defined.

Based on studies in rats, ascorbic acid is thought to lower plasma and liver copper levels by reducing copper absorption, and the reduced copper absorption later stimulates copper absorption and depresses biliary excretion (Van den Berg et al. 1994). The decrease in absorption is caused by a reduction of cupric (Cu^{2+}) ions to cuprous ions (Cu^{1+}), which are less well-absorbed (Lönnerdal 1996). High levels of ascorbic acid might also decrease ceruloplasmin oxidase activity, although the overall effect of ascorbic acid on absorption and metabolism of copper in humans may be less than in animals (Lönnerdal 1996; Turnlund 1988). Administration of ascorbic acid with zinc at 1 g per day in patients with Wilson disease showed no interaction of ascorbic acid and zinc, on copper balance, or ^{64}Cu absorption (Brewer et al. 1993a). Ascorbic acid had no detectable effect on the efficacy of zinc for copper-balance control in those patients.

Amino acids such as histidine and cysteine reduce copper absorption by forming complexes that are not well absorbed (Lönnerdal 1996). Histidine also enhances the inhibitory effects of zinc on copper absorption in rats (Wapnir and Balkman 1991). On the other hand, complexes of copper with other amino acids and organic acids might result in similar bioavailability to that of soluble copper sulfate (Wapnir 1998).

Based on studies to date, zinc is the primary mineral, and dietary element, which affects copper absorption. The effect of excess zinc on reducing copper absorption is well documented in a number of species (see summaries in Turnlund 1998; Wapnir 1998), and zinc has been used effectively in the treatment of Wilson disease (Hoogenraad et al. 1978; Hoogenraad et al. 1978; Hoogenraad et al. 1987; Brewer and Yuzbasiyan-Gurkan 1992; Brewer et al. 1994; Brewer et al. 1998). In pregnant rats, the consumption of diets containing high levels of zinc can result in fetal copper deficiency (Reinstein et al. 1984). In humans, the consumption of zinc supplements (50 mg per day for 6-8 weeks) can result in reductions in erythrocyte copper-zinc superoxide dismutase activity (Fisher et al. 1984; Yadrick et al. 1989; Johnson et al. 1998), suggesting that the chronic consumption of this level of zinc supplement could result in a condition of marginal copper status. Given the above possibility, the Institute of Medicine recommended that, at least for pregnant women, copper supplements (2 mg) should be provided to women when zinc supplements (25 mg or more) are given (IOM 1990).

In addition to zinc, various other minerals, such as iron, tin, calcium, phosphorus, cadmium, and molybdenum, interact with copper absorption and metabolism, although their effect compared with that of zinc is relatively minor or debatable in humans (Lönnerdal 1996; Turnlund 1988; Wapnir 1998). These minerals are cations that might compete for uptake in the digestive tract, thereby reducing absorption. These minerals might also affect copper utilization and excretion. For example, molybdenum has long been known to result in copper deficiency in ruminants but has little effect in nonruminants. Along with zinc for maintenance therapy, tetrathiomolybdate is now being used in the initial treatment of patients with the neurological or psychiatric form of Wilson disease. Tetrathiomolybate acts by blocking absorption of copper when given with food and by complexing with serum copper when given separately from food (Brewer et al. 1996).

Similar to other metals, the bioavailability of copper in soils or suspended particulates in water is likely to be a function of its mineral or surface sorbed form, solubility, and particle size (Davis et al. 1993; Ruby et al. 1996). As demonstrated for lead, solubility and bioavailability can vary greatly, depending on chemical and physical form (Ruby et al. in press). Copper acetate and sulfate are considerably more soluble and thus more bioavailable than cupric oxide (Johnson et al. 1998; Wapnir 1998), copper sulfides (e.g., chalcopyrite), and other less-soluble minerals. Copper in soil and sediments also adsorbs strongly to soil components, such as clay minerals, hydrous iron, and manganese oxides (Tyler and McBride 1982), resulting in reduced solubility and mobility.

CONCLUSIONS

- Copper is an essential nutrient.
- Studies of absorption, transport and metabolism of copper have provided insights into the biochemical mechanisms for coping with copper deficiency and excess.
- The retention of copper from the diet is influenced by age, amount and form of copper in the diet, and genetic background.
- Bioavailability of copper varies with age and diet composition.
- The liver plays a central role in copper homeostasis by varying the excretion of copper into the bile for loss in the stool.
- The newly discovered chaperones for copper have provided insight into how copper ions in cells are guided to their target proteins.

RECOMMENDATIONS

- Studies are needed to elucidate mechanisms of copper absorption, distribution, and excretion in humans.
- Studies are needed on age-related changes in copper absorption or retention.
- Research should be conducted on the genetic basis of the absorption mechanism and on whether variation in absorption efficiency has a genetic basis.
- Research is needed to define extrahepatic processes for uptake and distribution.
- The ability of copper to induce proteins involved in its metabolism and transport should be investigated. Particular emphasis should be given to the investigation of metal response elements on copper transport proteins.
- Research is needed to determine the ontogeny of copper transporters.
- Research is needed to identify how the form of copper (i.e., valence state and complexed forms) influences absorption and distribution.

REFERENCES

Amaravadi, R., D.M. Glerum and A. Tzagoloff. 1997. Isolation of a cDNA encoding the human homolog of COX17, a yeast gene essential for mitochondrial copper recruitment. Hum. Genet. 99(3):329-333.

Alda, J.O. and R. Garay. 1990. Chloride (or bicarbonate)-dependent copper uptake through the anion exchanger in human blood cells. Am. J. Physiol. 259(4 Pt 1):C570-C576.

Baker, D.H., J. Odle, M.A. Funk, and T.M. Wieland. 1991. Research note: bioavailability of copper in cupric oxide, cuprous oxide, and in a copper-lysine complex. Poult. Sci. 70(1):177-179.

Bremner, I. 1980. Absorption, transport and distribution of copper. Pp. 23-48 in Biological Roles of Copper. Ciba Foundation Symposium 79. Amsterdam: Excerpta Medica.

Brewer, G.J. and V. Yuzbasiyan-Gurkan. 1992. Wilson disease. Medicine 71(3):139-164.

Brewer, G.J., R.D. Dick, V. Yuzbasiyan-Gurkan, V. Johnson and Y. Wang. 1994. Treatment of Wilson's disease with zinc: XIII. Therapy with zinc in presymptomatic patients from the time of diagnosis. J. Lab. Clin. Med. 123(6):849-858.

Brewer, G.J., R.D. Dick, V.D. Johnson, J.A. Brunberg, K.J. Kluin, and J.K. Fink. 1998. Treatment of Wilson's disease with zinc: XV. Long-term follow-up studies. J. Lab. Clin. Med. 132(4):264-278.

Brewer, G.J., V. Johnson, R.D. Dick, K.J. Kluin, J.K. Fink, and J.A. Brunberg. 1996. Treatment of Wilson Disease with ammonium tetrathiomolybdate. II. Initial therapy in 33 neurologically affected patients and follow-up with zinc therapy. Arch. Neurol. 53(10):1017-1025.

Brewer, G.J., V. Yuzbasiyan-Gurkan, V. Johnson, R.D. Dick, and Y. Wang. 1993a. Treatment of Wilson's disease with zinc: XI. Interaction with other anticopper agents. J. Am. Coll. Nutr. 12(1):26-30.

Brewer, G.J., V. Yuzbasiyan-Gurkan, R. Dick, Y. Wang, and V. Johnson. 1993b. Does a vegetarian diet control Wilson's disease? J. Am. Coll. Nutr. 12(5):527-530.

Cordano, A. 1998. Clinical manifestations of nutritional copper deficiency in infants and children. Am. J. Clin. Nutr. 67(5 Suppl.):1012S-1016S.

Cousins, R.J. 1985. Absorption, transport, and hepatic metabolism of copper and zinc: Special reference to metallothionein and ceruloplasmin. Physiol. Rev. 65(2): 238-309.

Culotta, V.C., L.W. Klomp, J. Strain, R.L. Casareno, B. Krems, and Gitlin JD. 1997. The copper chaperone for superoxide dismutase. J. Biol. Chem. 272(38):23469-23472.

Culotta, V.C., S.J. Lin, P. Schmidt, L.W. Klomp, R.L. Casareno, and J. Gitlin. 1999. Intracellular pathways of copper trafficking in yeast and humans. Adv. Exp. Med. Biol. 448:247-254.

Dancis, A., D.S. Yuan, D. Haile, C. Askwith, D. Eide, C. Moehle, J. Kaplan, and R.D. Klausner. 1994a. Molecular characterization of a copper transport protein in *S. cerevisiae*: An unexpected role for copper in iron transport. Cell 76:393-402.

Dancis, A., D. Haile, D.S Yuan, and R.D. Klausner. 1994b. The *Saccharomyces cerevisiae* copper transport protein (Ctr1p). Biochemical characterization, regulation by copper, and physiologic role in copper uptake. J. Biol. Chem. 269(41):25660-25667.

Davis, A., J.W. Drexier JW, M.V. Ruby, and A. Nicholson. 1993. Micromineralogy of mine wastes in relation to lead bioavailability, Butte, Montana. Environ. Sci. Technol. 27(7):1415-1425.

Deneke, S.M. and B.L. Fanburg. 1989. Regulation of cellular glutathione. Am. J. Physiol. 257(4 Pt 1):L163-L173.

DiSilvestro, R.A., and R.J. Cousins. 1983. Physiological ligands for copper and zinc. Ann. Rev. Nutr. 3:261-288.

Dunn, M.A., M.H. Green and R.M. Leach, Jr. 1991. Kinetics of copper metabolism in rats: a compartmental model. Am. J. Physiol. 261(1 Pt 1):E115-25.

Eide, D.J. 1998. The molecular biology of metal ion transport in *Saccharomyces cerevisiae*. Ann. Rev. Nutr. 18:441-469.

Fields, M., J. Holbrook, D. Scholfield, J.C. Smith,Jr, and S. Reiser. 1986. Effect of fructose or starch on copper-67 absorption and excretion by the rat. J. Nutr. 116(4):625-632.

Fischer, P.W., A. Giroux, and M.R. L'Abbe. 1984. Effect of zinc supplementation on copper status in adult man. Am. J. Clin. Nutr. 40(4): 743-746.

Freedman, J.H., and J. Peisach. 1989a. Resistance of cultured hepatoma cells to copper toxicity. Purification and characterization of the hepatoma metallothionein. Biochim. Biophys. Acta 992(2):145-54.

Freedman, J.H. and J. Peisach. 1989b. Intracellular copper transport in cultured hepatoma cells. Biochem. Biophys. Res. Commun. 164(1):134-140.

Fuentealba, I., S. Haywood, and J. Trafford. 1989. Variations in the intralobular distribution of copper in the livers of copper-loaded rats in relation to the pathogenesis of copper storage diseases. J. Comp. Pathol. 100(1):1-11.

Georgatsou, E., L.A. Mavrogiannis, G.S. Fragiadakis, and D. Alexandraki. 1997. The yeast Fre1p/Fre2p cupric reductases facilitate copper uptake and are regulated by the copper-modulated Mac1p activator. J. Biol. Chem. 272:13786-13792.

Gross, J.B., Jr., B.M. Myers, L.J. Kost, S.M. Kuntz, and N.F. LaRusso. 1989. Biliary copper excretion by hepatocyte lysosomes in the rat. Major excretory pathway in experimental copper overload. J. Clin. Invest. 83(1):30-39.

Harris, E.D. 1995. Role of ligands in the translocation of metals. Pp. 71-88 in Handbook of Metal-Ligand Interactions in Biological Fluids. Bioinorganic Medicine, Vol. 1, G. Berthon, ed. New York: Marcel Dekker.
Harris, E.D. 1997. Copper. Pp. 231-273 in Handbook of Nutritionally Essential Mineral Elements, B.L. O'Dell and R.A. Sunde, eds. New York: Marcel Dekker.
Harris, D.I. and A. Sass-Kortsak. 1967. The influence of amino acids on copper uptake by rat liver slices. J. Clin. Invest. 46(4):659-677.
Harris, Z.L., Y. Takahashi, H. Miyajima, M. Serizawa, R.T. MacGillivray and J.D. Gitlin. 1995. Aceruloplasminemia: molecular characterization of this disorder of iron metabolism. Proc. Natl. Acad. Sci. (USA) 92(7):2539-2543.
Hassett, R. and D.J. Kosman. 1995. Evidence for Cu(II) reduction as a component of copper uptake by Saccharomyces cerevisiae. J. Biol. Chem. 270(1):128-134.
Haywood, S. 1981. A non-random distribution of copper within the liver of rats. Br. J. Nutr. 45(2):295-300.
Hillman, L.S., L. Martin and B. Fiore. 1981. Effect of oral copper supplementation on serum copper and ceruloplasmin concentrations in premature infants. J. Pediatr. 98(2):311-313.
Hogan, G.D. and W.E. Rauser. 1981. Role of copper binding, absorption, and translocation in copper tolerance of agrostis gigantea roth. J. Exp. Bot. 32(126):27-36.
Holt, D., D. Dinsdale, and M. Webb. 1986. Intestinal uptake and retention of copper in the suckling rat, *Rattus rattus*. I. Distribution and binding. Comp. Biochem. Physiol. 83(2):313-316.
Holtzman, N.A. and B.M. Gaumnitz. 1970. Studies on the rate of release and turnover of ceruloplasmin and apoceruloplasmin in rat plasma. J. Biol. Chem. 245(9):2354-2358.
Hoogenraad, T.U., R. Koevoet, and E.G. de Ruyter Korver. 1979. Oral zinc sulphate as long-term treatment in Wilson's disease (hepatolenticular degeneration). Eur. Neurol. 18(3):205-211.
Hoogenraad, T.U., C.J. van den Hamer, R. Koevoet, and E.G. Korver. 1978. Oral zinc in Wilson's disease [letter]. Lancet 2(8102):1262-1263.
Hoogenraad, T.U., J. van Hattum, C.J. van den Hamer. 1987. Management of Wilson's disease with zinc sulfate. Experience in a series of 27 patients. J. Neurol. Sci. 77(2-3):137-146.
Houwen, R., M. Dijkstra, F. Kuipers, E.P. Smit, R. Havinga and R.J. Vonk. 1990. Two pathways for biliary copper excretion in the rat. The role of glutathione. Biochem. Pharmacol. 39(6):1039-1044.
Humbert W., M. Aprahamian, C. Stock, and J.F. Grenier. 1982. Copper accumulation in primary biliary cirrhosis. An electron and X-ray microanalytical study. Histochemistry 74(1):85-93.

Irie S, and M. Tavassoli. 1986. Liver endothelium desialates ceruloplasmin. Biochem. Biophys. Res. Commun. 140(1):94-100.
IOM (Institute of Medicine). 1990. Nutrition during pregnancy. Part I: Weight Gain. Part II: Nutrient Supplements. Washington, DC: National Academy Press.
Iyengar V, G.J. Brewer, R.D. Dick, and O.Y. Chung. 1988. Studies of cholescystokinin-stimulated biliary secretions reveal a high molecular weight copper-binding substance in normal subjects that is absent in patients with Wilson's disease. J. Lab. Clin. Med. 111(3):267-274.
Johnson, M.A., M.M. Smith, and J.T. Edmonds. 1998. Copper, iron, zinc, and manganese in dietary supplements, infant formulas, and ready-to-eat breakfast cereals. Am. J. Clin. Nutr. 67(5 Suppl.):1035S-1040S.
L'Abbe M.R., and P.W.R. Fischer. 1984. The effects of dietary zinc on the activity of copper-requiring metalloenzymes in the rat. J. Nutr. 114(5):823-828.
Linder M. and C.A. Goode. 1991. Biochemistry of Copper. New York: Plenum Press.
Lönnerdal, B. 1996. Bioavailability of copper. Am. J. Clin. Nutr. 63(5 Suppl.):821S-829S.
Lönnerdal, B. 1998. Copper nutrition during infancy and childhood. Am. J. Clin. Nutr. 67(5 Suppl.):1046S-1053S.
Lönnerdal, B., J.G. Bell, and C.L. Keen. 1985. Copper absorption from human milk, cow's milk, and infant formulas using a suckling rat model. Am. J. Clin. Nutr. 42(5):836-844.
Murata Y, E. Yamakawa, T. Lizuka, H. Kodama, T. Abe, Y. Seki, and M. Kodama. 1995. Failure of copper incorporation into ceruloplasmin in the Golgi apparatus of LEC rat hepatocytes. Biochem. Biophys. Res. Commun. 209:349-355.
NRC (National Research Council). 1989. Recommended Dietary Allowances, 10th Ed. Washington, D.C.: National Academy Press.
Oestreicher P, and R.J. Cousins. 1985. Copper and zinc absorption in the rat: Mechanism of mutual antagonism. J. Nutr. 115(2):159-166.
Percival, S.S. 1998. Copper and immunity. Am. J. Clin. Nutr. 67(5 Suppl.):1064S-1068S.
Percival S.S., and E.D. Harris. 1989. Ascorbate enhances copper transport from ceruloplasmin into human K562 cells. J. Nutr. 119(5):779-784.
Percival S.S., and E.D. Harris. 1990. Copper transport from ceruloplasmin: Characterization of the cellular uptake mechanism. Am. J. Physiol. 258(1 Pt 1):C140-C146.
Percival S.S., and E.D. Harris. 1991. Regulation of Cu, Zn superoxide dismutase with copper. Caeruloplasmin maintains levels of functional enzyme activity during differentiation of K562 cells. Biochem. J. 274(Pt 1):153-158.
Portnoy M.E., A.C. Rosenzweig, T. Rae, D.L. Huffman, T.V. O'Halloran,

and V.C. Culotta. 1999. Structure-function analyses of the ATX1 metallochaperone. J. Biol. Chem. 274(21):15041-15045.

Reinstein, N.H., B. Lonnerdal, C.L. Keen and L.S. Hurley. 1984. Zinc-copper interactions in the pregnant rat: fetal outcome and maternal and fetal zinc, copper and iron. J. Nutr. 114(7):1266-1279.

Rolfs A, and M.A. Hediger. 1999. Metal ion transporters in mammals: structure, function and pathological implications. J. Physiol. 518 (Pt 1):1-12.

Ruby M.V., A. Davis, R. Schoof, S. Eberle, and C.M. Sellstone. 1996. Estimation of lead and arsenic bioavailability using a physiologically based extraction test. Environ. Sci. Technol. 30(2):422-430.

Ruby M.V., R. Schoof, J. Drexier, W. Brattin, M. Goldade, G. Post, M. Harnois, W. Berti, M. Carpenter, D. Edwards, D. Cragin, and W. Chappell. In press. Advances in evaluating the oral bioavailbity of inorganics in soil for use in human health risk assessment. Environ. Sci. Technol.

Sann, L., D. Rigal, G. Galy, F. Benvenu and J. Bourgeois. 1980. Serum copper and zinc concentration in premature and small-for-date infants. Pediatr. Res. 14(9):1040-1046.

Schmitt R.C., H.M. Darwish, J.C. Cheney, and M.J. Ettinger. 1983. Copper transport kinetics by isolated rat hepatocytes. Am. J. Physiol. 244:G183-G191.

Srikumar T.S., G.K. Johansson, P.A. Ockerman, J.A. Gustafsson, and B. Akesson. 1992. Trace element status in health subjects switching from mixed to a lactovegetarian diet for 12 mo. Am. J. Clin. Nutr. 55(4):885-890.

Sternlieb, I. 1980. Copper and the liver. Gastroenterology 78(6):1615-1628.

Tong, K.K., and H.J. McArdle. 1995. Copper uptake by cultured trophoblast cells isolated from human term placenta. Biochim. Biophys. Acta 1269(3):233-236.

Turnlund, J.R. 1988. Copper nutriture, bioavailability, and the influence of dietary factors. J. Am. Diet. Assoc. 88(3):303-308.

Turnlund, J.R. 1998. Human whole-body copper metabolism. Am. J. Clin. Nutr. 67(5 Suppl.):960S-964S

Turnlund, J.R., W.R. Keyes, G.L. Pfeiffer, and K.C. Scott. 1998. Copper absorption, excretion, and retention by young men consuming low dietary copper determined by using the stable isotope $^{65}Cu^{1,2}$. Am. J. Clin. Nutr. 67(6):1219-1225.

Tyler L.D., and M.B. McBride. 1982. Mobility and extractability of cadmium, copper, nickel, and zinc in organic and mineral soil columns. Soil Sci. 134(3):198-205.

Uauy, R., M. Olivares and M. Gonzalez. 1998. Essentiality of copper in humans. Am. J. Clin. Nutr. 67(5 Suppl.):952S-959S.

Valentine, J.S. and E.B. Gralla. 1997. Delivering copper inside yeast and human cells. Science 278(5339):817-818.

van den Berg, G.J., and H.J. McArdle. 1994. A plasma membrane NADH oxidase is involved in copper uptake by plasma membrane vesicles isolated from rat liver. Biochim. Biophys. Acta 1195(2):276-280.

van den Berg, G.J., S. Yu, A.G. Lemmens, and A.C. Beynen. 1994. Ascorbic acid feeding of rats reduces copper absorption, causing impaired copper status and depressed biliary copper excretion. Biol. Trace Elem. Res. 41(1-2):47-58.

Wapnir, R.A. 1998. Copper absorption and bioavailability. Am. J. Clin. Nutr. 67(5 Suppl.):1054S-1060S.

Wapnir R.A., and C. Balkman. 1991. Inhibition of copper absorption by zinc: Effect of histidine. Biol. Trace Elem. Res. 29(3):193-202.

WHO (World Health Organization). 1996. Trace Elements in Human Nutrition and Health. Geneva: World Health Organization.

Widdowson, E.M. 1974. Trace elements in foetal and early postnatal development. Proc. Nutr. Soc. 33(3):275-84.

Yadrick, M.K., M.A.Kenney, and E.A. Winterfeldt. 1989. Iron, copper, and zinc status: response to supplementation with zinc or zinc and iron in adult females. Am. J. Clin. Nutr. 49(1):145-50.

Zhou B., and J. Gitschier. 1997. *hCTR1*: A human gene for copper uptake identified by complementation in yeast. Proc. Natl. Acad. Sci. (USA) 94:7481-7486.

3

Health Effects of Copper Deficiencies

THE essentiality of copper for animals was reported in 1928 in a study showing that it is essential for erythropoiesis in rats fed a milk-based diet (Hart et al. 1928). Erythropoiesis was improved when copper sulfide containing ash was added back to the diet (Hart et al. 1928). Reports of copper-deficiency in grazing livestock followed, further substantiating the essentiality of copper (Neal et al. 1931). In humans, the essentiality of copper was clearly demonstrated by studies showing anemia, neutropenia, and bone-marrow abnormalities in young children with copper deficiencies (Cordano et al. 1964). The children were responsive to copper therapy. Additional studies have demonstrated the essentiality of copper in immune function, bone formation, red- and white-blood-cell maturation, lipid metabolism, iron transport, myocardial contraction, and neurological development (Danks 1988).

TERATOGENESIS OF COPPER DEFICIENCY

Causes of Copper Deficiency

Before a discussion of the developmental effects associated with a deficit of copper, it is important to recognize the multiple ways an embryonic or fetal copper deficiency might arise. First, a deficiency can occur if the mother has a low dietary intake of copper (primary deficiency). An insufficient intake of copper in the diet (combined intake from solids and liquids) will eventually result in a primary deficiency of copper and potentially death.

A secondary, or conditioned, deficiency might occur even if the mother's intake of copper from the diet is adequate. Conditioned deficiencies can arise by several means. First, copper deficiency can arise through an effect of drugs or other chemicals on the metabolism of copper. Second, a conditioned embryonic or fetal deficiency of copper might arise if the mother has an excessive low intake of copper as a consequence of an underlying disease or if disease-induced changes in maternal copper metabolism reduces the transfer of copper to the conceptus. Third, nutritional interactions can produce conditioned deficiencies. These interactions can be of several types. For example, copper-binding factors, such as phytate and possible fiber, in the mother's diet can potentially reduce the amount of copper absorbed from the diet. A copper deficiency can also occur if the diet contains a high concentration of a metal with physical-chemical properties similar to those of copper; zinc, cadmium and silver are examples of metals in this category. Finally, a conditioned copper deficiency can occur as a consequence of genetic factors. Those can either be a single gene defect (e.g., Menkes disease) or multiple genes that collectively affect one or more aspects of copper metabolism. In experimental animals, multiple gene effects are typically referred to as a strain or breed effect. Those effects are discussed in more detail below.

Copper in Prenatal Development

The importance of copper for the prenatal development of mammals was shown in sheep by Bennetts et al. (1948) in their demonstration that enzootic ataxia, a disease affecting the developing fetus, could be prevented by giving the ewe additional copper during pregnancy. This disorder is characterized by spastic paralysis, especially of the hind limbs, severe incoordination, blindness in some cases, and anemia. Typically, the brains of animals with enzootic ataxia are small and characterized by collapsed cerebral hemisphoresis, shallow convolutions, and a paucity of normal myelin (Hurley and Keen 1979). Similar neonatal ataxia and brain abnormalities have been reported in newborn copper-deficient deer, goats, swine, guinea pigs, and rats (Hurley and Keen 1979, Yoshikawa et al. 1996). Copper deficiency, as evidenced by low concentrations of copper in plasma, can be induced in most mammalian species by feeding a copper-deficient diet (copper at less than 1 mg/kg of body weight compared with control diets of 8 to 15 mg/kg) for 2 to 4 weeks.

Researchers have not agreed on the biochemical bases of the brain abnormalities associated with copper deficiency during early development.

One possibility is a reduction in the activity of the cuproenzyme cytochrome c oxidase. Mills and Williams (1962) found that, when copper is deficient, the activity of cytochrome c oxidase is significantly reduced in the large motor neurons of the red nucleus of the brain, an area where degeneration is often striking. It can be argued that sufficient reduction in cytochrome c oxidase activity causes cellular anoxia, resulting in tissue death. Inadequate production of ATP for normal phospholipid synthesis might explain in part the high amount of amyelination typically observed in brains of copper-deficient fetuses and neonates (Hurley and Keen 1979).

Although the functional significance of copper-deficiency-induced reductions in cytochrome c activity in the brain is not clear, such reductions in the liver and heart are associated with reduced ATP production (Weisenberg et al. 1980; Kuznetsov et al. 1996). Copper-deficiency-induced reductions in cytochrome c oxidase are in part due to low amounts of the assembled protein aa3 (Rossi et al. 1998). Cell-culture experiments show that copper-deprived cells with low cytochrome c oxidase activity can have increased manganese superoxide dismutase (MnSOD) activity, indicating increased mitochondrial oxidative stress. Moreover, the cells have increased protein carbonyls, an indicator of protein oxidation. Thus, the copper-deprived state might result in oxidative damage to respiratory chain proteins, such as complex I (Johnson and Thomas, 1999). Reduction in cytochrome c oxidase might also result in an increased leakage of electrons through the electron transport chain to molecular oxygen, thereby increasing production of reactive oxygen species (ROS) (Fantel 1996).

Excessive cellular oxidative damage might be a second mechanism contributing to the developmental abnormalities associated with copper deficiency. Copper-zinc superoxide dismutase (CuZnSOD) in the brain was reported to be low in young copper-deficient rats (Prohaska and Wells 1975). Reductions in CuZnSOD activity are associated with excessive lipid and protein oxidative damage and cell death. Specifically, the downregulation of CuZnSOD results in neuronal-cell death (Troy and Shelanski 1994). Similarly, mutations to the human CuZnSOD gene are linked to degeneration of the motor neurons (Gurney et al. 1994). These studies have found an increase in hydroxyl radical generating capacity in *SOD1* mutants (a gene that codes for analogous proteins in yeast). The increase is likely promoted by the release of copper from the mutant enzyme, which can enhance Fenton reaction activity (Eum and Kang 1999). In addition to promoting injury of neuronal cells, reductions in CuZnSOD activity result in brain edema and focal cerebral ischemic injury in CuZnSOD mutant mice (Kondo et al. 1997). Similarly, MnSOD null mice suffer from neurological pathologies that are likely due to oxidative-stress-induced damage (Melov et al. 1998). Prion protein (PrP^c), a glycoprotein normally expressed in neurons, might play a role in regulating CuZnSOD activity

(Brown et al. 1997). Copper stimulates endocytosis of PrP^c from the cell surface, and the PrP^c might act as a recycling receptor for copper ion uptake (Pauly and Harris 1998). Cerebellar cells of PrP^c null mice (Prnp$^{0/0}$) have low CuZnSOD activity and are particularly susceptible to oxidative stress.

Despite low CuZnSOD activity, the peroxidation of brain lipids in young copper-deficient rats does not appear to be excessive (Prohaska and Wells 1975). The lack of excessive lipid peroxidation can be explained by a shift in the fatty-acid profile to a less peroxidizable composition in the brains of copper-deficient rats. Alterations in the fatty-acid composition of select tissues in response to oxidative insults in adult tissues have been reported (Zidenberg-Cherr and Keen 1991). Although the type or amount of embryonic polyunsaturated fatty acid influenced by increased oxidative pressures has not been determined, a change in the fatty-acid composition to a more saturated profile could result in dysmorphogenesis. However, in in vitro embryo culture models, the teratogenic effects of copper deficiency are reduced if the medium is supplemented with antioxidant enzymes (Hawk et al. 1995).

Copper deficiency can also affect several other brain cuproenzymes. For example, the activity of peptidylglycine α-amidating monooxygenase (PAM) (EC 1.14.17.3) is markedly reduced with developmental copper deficiency (Prohaska and Bailey 1995). PAM is responsible for converting a number of precursors into their α-amide form, including gastrin, cholecystokinin, oxytocin, vasopressin, neuropeptide Y, and vasoactive intestinal peptide; therefore, a reduction in PAM activity might result in altered physiological functioning (Prohaska and Bailey 1995). PAM activity can remain low even after months of copper repletion (Prohaska and Bailey 1995). It has not been determined whether the persistence of low PAM activity in the brain after prenatal copper deficiency is due to the relatively long period of time that it takes to increase brain copper concentrations or whether it represents a more fundamental epigenetic defect. In either case, given PAM's role in hormone activation, the duration of reduced PAM activities in the brain could have a significant impact on long-term behavioral and metabolic events.

The extent to which copper-deficiency-induced changes in the activities of brain cuproenzymes contribute to the morphological damage in fetal and neonatal brains is unknown. It is reasonable to postulate, however, that copper-deficiency-induced developmental alterations in the activities of CuZnSOD, cytochrome c oxidase, and PAM are contributing factors to persistent behavioral and functional defects associated with prenatal copper deficiency (Hunt and Idso 1995; Prohaska and Hoffman 1996).

In addition to brain defects, copper-deficient fetuses and neonates are typically characterized by severe connective tissue abnormalities. Cardiac hemorrhages are a frequent finding in copper-deficient sheep, rats, guinea pigs, and mice (Hurley and Keen 1979). The walls of the internal and common carotid arteries in copper deficient-fetuses tend to have an endothelium that appears normal but has sparse, poorly developed elastin. Cerebral arteries are also often characterized by a low elastin content. Furthermore, the elastin that is present does not have the concise fibrillar arrangements seen in control animals. The reduction in elastin content and cross-linking integrity is primarily due to a decrease in the activity of the cuproenzyme lysyl oxidase, which catalyzes the oxidation of certain peptidyl lysine and hydroxylysine residues to peptide aldehydes, initiating the cross-linking mechanisms required for connective-tissue stability (Rucker et al. 1998).

In addition, it is important to note that copper is associated with factors located in the extracellular matrix, which has angiogenic properties. Specifically, SPARC (secreted protein, acidic, rich in cysteine), an extracellular matrix protein involved in the regulation of endothelial proliferation, has an active copper-binding domain (Lane et al. 1994). The second cation region of SPARC is copper binding and stimulates angiogenesis (Lane et al. 1994). Given the angiogenic properties of copper and low-molecular-weight copper complexes, it is reasonable to suggest that altered angiogenesis can contribute to the brain dysmorphology associated with developmental copper deficiency. Finally, copper deficiency has also been associated with an acceleration of proteolysis of collagen and elastin due to nonspecific proteases that migrate from the blood compartment into vascular tissue (Romero et al. 1989). It is evident that copper deficiency can result in abnormal vascular development through a variety of mechanisms.

Skeletal defects can occur as a result of copper deficiency. Lambs with enzootic ataxia typically have poorly developed, light, brittle bones that fracture frequently. Bone abnormalities have been found in copper-deficient calves and fowls. In dogs and swine, the young born to females fed copper-deficient diets had deformed leg bones (Hurley and Keen 1979). The lesions appeared to be associated with an impairment of osteogenesis, resulting in thinning of the cortex and trabeculae of the long bones. Copper-deficient chicks are different, having severe hypoplasia of the long bones and leg weakness. Amine oxidase, cytochrome oxidase, and lysyl oxidase activities are low, and the ratio of soluble-to-insoluble collagen is high. The increased fragility of copper-deficient bones appears to result from the low number of cross-links present in the collagenous matrix (Rucker et al., 1998).

Finally, lung abnormalities are a frequent consequence of prenatal and early postnatal copper deficiency. Lungs from neonatal rabbits born to dams fed copper-deficient diets were characterized by low concentrations of copper and lysyl oxidase activity and high proportions of poorly cross-linked elastin and collagen. The lungs were also characterized by low concentrations of surfactant phospholipids (Abdel-Mageed et al. 1994). Similar results have been reported in copper-deficient rats (Dubick et al. 1985).

Drug-Induced Copper Deficiency

Rosa (1986) reviewed a series of cases of infants born to women who received d-penicillamine (DPA) during pregnancy for a variety of conditions, including connective-tissue abnormalities and rheumatoid arthritis. Abnormalities observed in the infants included lax skin, hyperflexibility of the joints, fragility of the veins, and numerous soft tissue abnormalities. It was suggested that the malformations were in part due to a drug-induced copper deficiency during embryonic or fetal development. Similar abnormalities were produced with DPA in rodent models (Keen et al., 1983b), and the teratogenicity of the drug can be modulated by maternal dietary copper intake (Mark-Savage et al., 1983). The above suggest that in humans maternal copper status should be monitored when DPA is taken during pregnancy.

Little is known about the influence of other copper chelating drugs such as captopril, disulfiram, dimercaptosuccinic acid (DMSA), and triethylenetetramine (TETA) on human prenatal development. Given the significant teratogenic effects that can be associated with similar drugs in experimental animals, it is reasonable to suggest that these drugs can pose a substantial risk to the conceptus if maternal dietary copper intake is low.

A condition of severe copper deficiency can be rapidly induced in experimental animals through the use of a number of chelating drugs, including disulfiram, DPA, TETA, and DMSA (Salgo and Oster 1974; Keen et al. 1983a,b; Taubeneck et al. 1992; Jasim et al. 1985). Each of those drugs is known to be teratogenic. The abnormalities produced are reminiscent of those induced by dietary copper deficiency. Although the teratogenicity of DPA, TETA, and DMSA can be modulated by the amount of copper in the mother's diet (Cohen et al. 1983; Mark-Savage et al. 1983), it is important to note that drugs that bind copper typically also bind zinc. Thus, the teratogenic effects of those drugs might be due to a combination of copper and zinc deficiencies.

In contrast to the effects of zinc, drugs and chemicals that induce an acute-phase response in the mother do not necessarily influence copper uptake by the embryo or fetus, and in most cases, fetal copper concentrations are unaffected by chemicals that induce transitory acute-phase responses (Keen 1996). That is not surprising, because maternal plasma copper concentrations are increased during an acute-phase response (due to the hepatic production and release of the cuproenzyme ceruloplasmin), and zinc concentrations are decreased. Ceruloplasmin has been postulated to be involved directly in copper transport to the embryo and fetus (Lee et al. 1993). It must be noted that the mechanisms underlying copper transport into the embryo and fetus is an area of active research.

Disease-Induced Copper Deficiency

Copper deficiency can occur secondarily to such diseases as chronic diarrhea, diabetes, alcoholism, and hypertension (Rosa 1986; Dubick et al. 1987; Turnlund 1994). In the case of maternal diabetes and alcoholism, disease-induced deficiencies of copper in the embryo or fetus have been postulated to contribute to the teratogenesis (Uriu-Hare et al. 1989; Zidenberg-Cherr et al. 1988; Dubick et al. 1999). Mothers with diabetes are reported to have low concentrations of copper in erythrocytes (Speich et al. 1992). In addition, maternal hypocupremia is often noted in cases of spontaneous abortion or rupture of the fetal and placental membranes (Kiilholma et al. 1984; Artal et al. 1979).

Copper-Diet Interactions

To date, there have been no reports of zinc-induced copper deficiencies affecting human pregnancy outcomes. However, zinc-induced fetal copper deficiencies are relatively easy to produce in experimental animals, and care should be taken to monitor women who consume zinc supplements during pregnancy. The Institute of Medicine recommends that copper supplements (2 mg) be provided to women taking zinc supplements (25 mg) during their pregnancies (IOM 1990).

In ruminants, copper deficiency can also be induced by the feeding of high concentrations of molybdenum, which, along with sulfate, can form a complex with copper that limits its absorption. The copper deficiency induced in this manner can be sufficient to pose a developmental risk (Howell et al. 1993).

Shavlovski et al. (1995) reported that embryonic copper deficiency can occur as a consequence of maternal silver toxicity. The above is of particular interest given that the authors provided data suggesting that silver-induced fetal copper deficiency occurs as a consequence of silver-induced alterations in ceruloplasmin synthesis. Shavlovski et al. (1995) postulated that the observed fetal copper deficiency was a consequence of a reduction in ceruloplasmin-mediated placental copper transport. Although the above hypothesis is attractive, it is in sharp contrast to the observation of normal copper transport in patients with the genetic disorder aceruloplasminemia (see Chapter 4). One interpretation of the above is that when there are genetic defects in ceruloplasmin synthesis, the embryo might be able to expand alternative pathways for copper transport. For example, it could be argued that there is an increase in the amount of copper transported by copper ATPases. However, the ability to modulate those pathways might be lost by the fetal stage.

Gene-Induced Copper Deficiency

There are at least two genetic defects that are expressed as copper-deficiency syndromes (Menkes disease and occipital horn syndrome). Both disorders are due to defects in a copper-transporting P-type ATPase. Infants with Menkes disease are characterized by progressive degeneration of the brain and spinal cord, hypothermia, connective-tissue abnormalities, and failure to thrive. Menkes disease has been recognized as a disorder of copper metabolism for over 20 years. The prognosis for infants with the disorder is poor and death typically occurs before 3 years of age (Danks 1988; Turnlund 1994). Similar to the blotchy mouse, the developmental abnormalities associated with Menkes disease are thought to be the consequence of low activity of numerous cuproenzymes during embryonic and fetal development. Those cuproenzymes include dopamine-B-monoxygenase (DBH), PAM, cytochrome c oxidase, lysyl oxidase, and CuZnSOD (Kaler 1998; Prohaska et al. 1997; Medeiros and Wildman 1997; Mercer 1998). The aberrant pattern of the plasma and cerebrospinal fluid of Menkes patients has been attributed to low activity of DBH (Kaler 1998). Moreover, individuals with Menkes disease are characterized by low activity of ceruloplasmin and PAM (Prohaska et al. 1997). Thus, low activity of numerous enzymes that rely on the amidation of peptides for their activity might occur as a secondary effect of low PAM activity.

The development of connective tissue abnormalities, such as bladder diverticula and vascular tortuosity, is likely attributable to alterations in lysyl oxidase activity. Abnormalities in connective tissues of Menkes sufferers are overt and might be more pronounced because, unlike other copper-dependent enzymes that draw copper from cytoplasmic carriers. It is postulated that lysyl oxidase incorporates its copper from the ATP7A once it crosses into the trans-Golgi (Mercer 1998).

Even though Menkes disease and occipital horn syndrome are the result of mutations in the same gene, they differ in their expression. Menkes syndrome is expressed at birth, and individuals with occipital horn syndrome are reported to be phenotypically normal at birth. However, detailed biochemical studies of infants with occipital horn syndrome at birth have not been reported (Proud et al. 1996).

Aceruloplasminemia is a rare autosomal recessive genetic defect that results in a lack of holoceruloplasmin production and an alteration in iron metabolism (Yoshida et al. 1995; Harris et al. 1995). One form of the mutation has been characterized by a mutation in the ceruloplasmin gene. The mutation results in a condition in which there is essentially no holoceruloplasmin (Harris et al. 1998). Ceruloplasmin functions as a ferroxidase. A deficiency of ceruloplasmin results in the accumulation of iron in select tissues, including the brain and pancreas. Individuals suffering from hereditary ceruloplasmin deficiency often develop diabetes mellitus secondary to iron-induced pancreatic damage (Kato et al. 1997). The hematochromatosis-like symptoms of aceruloplasminemia include an ataxic gait, dysarthria, retinal degeneration, neuropathy, and diabetes mellitus (Miyajima et al. 1996; Takahashi et al. 1996). Although the specific effects of aceruloplasminemia on embryonic and fetal development have yet to be delineated, women with this defect are able to conceive and have normal pregnancies.

A number of strains of mice, rats, and sheep are characterized by abnormal copper metabolism, and for several of these strains, the abnormality can influence embryonic or fetal development.

Interactions between copper and genetic factors can be classified into two groups. The first type involves strain differences that produce a differential response to diets that are deficient or marginal in copper, and the second type involves a single mutant gene, the expression of which can resemble the signs of a deficiency or a toxicity of the element. That expression can be reduced or prevented by nutritional manipulation. Those two types of phenomena can interact. Thus, the phenotypic expression of a mutant gene can be modulated by the strain background.

Examples of the first type of gene-nutrient interaction are certain mutant genes in mice. For example, the mottled (Mo) mouse is characterized by a defect in cellular copper transport, which is phenotypically expressed by signs of copper deficiency (Mercer et al. 1994). Over 10 alleles at the mottled locus have been described. They range in severity from hypopigmentation of hair at birth to death in utero. The blotchy mutant (MoB) is typically characterized by severe connective-tissue defects and neurological abnormalities, similar to those observed with severe maternal copper deficiency. The primary genetic defect in the blotchy mouse is thought to involve a mutation in a copper-transporting ATPase gene that is homologous to the Menkes gene in humans (Das et al. 1995). The phenotypic expression is thought to result from the reduced activity of several cuproenzymes.

Representative of the second category of gene-nutrient interaction is the observed influence of the breed of sheep on the incidence of enzootic ataxia within a geographical area (Wiener et al. 1978). For example, the offspring of Welsh sheep often have a lower incidence of enzootic ataxia than do offspring of blackface sheep, even when the ewes are maintained on the same pasture. The difference in the occurrence of enzootic ataxia has been correlated with the mother's ability to absorb copper (Wiener et al. 1978).

Human Copper Deficiency and Teratogenesis

Primary Dietary Copper Deficiency

In the United States, the adult population typically has an intake of copper that is below the estimated safe and adequate dietary recommendation of 1.5-3.0 mg of copper per day (Milne 1998). Women of childbearing age are also unlikely to meet their putative copper requirement (NRC 1989). Although overt copper deficiency is not commonly seen in the United States, some authors argue that moderate copper deficiency is more prevalent (Danks 1988; Reiser et al. 1985; Kelley et al. 1995; Milne and Nielsen 1996). According to Klevay (1998), consumption of diets providing less than 1 mg of copper per day can be associated with adverse health effects. Copper deficiency occurs in a variety of conditions, including diabetes, hypertension, alcoholism, and total parental nutrition feedings (Tokuda et al. 1986; Danks 1988; Shaw 1992: Olivares and Uauy 1996; Uauy et al. 1998). There is considerable debate however, on the extent to which copper deficiency influences human prenatal development. Brewer et al. (2000) studied 26 pregnancies of 19 women with Wilson disease. All were treated with zinc as their sole anticopper drug. Of the 26 newborns, 24 were normal, one had a surgically correctable heart defect, and another had anencephaly. Buamah et al. (1984) reported that the finding of low serum copper concentrations in pregnant women during mid-gestation was a risk factor for anencephaly. Morton et al. (1976) reported a significant correlation between low copper content in drinking water and the occurrence of neural-tube defects in South Wales. Dietary copper intake was not considered, and the observation has not been confirmed.

HEALTH EFFECTS OF COPPER DEFICIENCIES IN ADULTS

Clinical copper deficiency in adults can occur, but it is rare. Data on clinical copper deficiency is to be differentiated from copper intake data, which infer a certain frequency of copper deficiency which may be higher than that which actually occurs. Clinical copper deficiency in the United States is also to be differentiated from hypotheses suggesting that low intake of copper leads to an increased risk of atherosclerosis. Clinical copper deficiency is most often due to zinc ingestion, which blocks copper absorption (see section on zinc treatment of Wilson disease). To produce copper deficiency, zinc ingestion has to occur over a period of time, probably a minimum of 2 months. Therefore, ingestion of zinc (e.g., to treat of colds with zinc lozenges) is not a risk if ingested for a few days only per episode. In addition, zinc taken with food tends to get bound to substances in food and has much less effect on copper absorption. Occasionally, zinc ingestion results from intentional swallowing of foreign objects that contain zinc and that remain in the stomach for long periods. More rarely, copper deficiency can result from surgical removal of a large section of the small intestine, thus greatly reducing absorption of copper.

The most sensitive indicator of clinical copper deficiency is the serum ceruloplasmin level. The synthesis and release of this copper-containing protein into the blood by the liver is dependent upon copper availability. As availability decreases, plasma ceruloplasmin decreases. The consequence of modest reductions in serum ceruloplasmin have not been defined.

As levels of copper, or copper availability, decrease further, there can be effects on the bone marrow. A sensitive cell line with regard to copper deficiency is the red cell line, perhaps because copper is required for heme synthesis. Thus, anemia is often thought to be an early bone-marrow effect. The more long-standing the copper deficiency the more likely the red cell indices will become hypochromic microcytic. Because copper is also required for cellular proliferation, the effect on the bone marrow can also produce leukopenia, particularly neutropenia, and thrombocytopenia (Cordano 1998).

In addition to its effect on the erythrocyte pool, an early effect of copper deficiency can be significant changes in the immune system (Percival 1998). In rats, an early sign of copper deficiency can be an impairment in the respiratory burst and candidacidal activity of macrophages (Babu and Failla 1990). Although the mechanisms that underlie the effects of copper deficiency on immune cells have not been defined, it is known that one early effect is alterations in interleukin 2 as well as interleukin 1 production (Bala and Failla 1992). The extent to which those effects occur in humans is not known.

Long-standing severe copper deficiency produces a neurological syndrome that appears to be primarily a peripheral neuropathy, in which loss of sensation and muscle weakness occurs.

CONCLUSIONS

- Severe deficiency of copper can have adverse developmental consequences.
- Copper deficiency can arise through a multitude of mechanisms, in-

cluding low dietary intake, genetic abnormalities, nutrient-nutrient interactions (e.g., copper and zinc), and nutrient-drug interactions.

• The frequency of copper deficiency or sufficiency in the United States has not been well defined for any age group.

• Clinical cases of copper deficiency in adults can occur as a result of poor absorption due to removal of a large section of the small intestine or excessive zinc intake.

• Marginal copper intake might have adverse health effects, on the vascular and immune system.

REFERENCES

Abdel-Mageed, A.B., R. Welti, F.W. Oehme, and J.A. Pickrell. 1994. Perinatal hypocuprosis affects synthesis and composition of neonatal lung collagen, elastin and surfactant. Am. J. Physiol. 267(6 Pt 1):L679-685.

Artal, R., R. Burgeson, F.J. Fernandez, and C.J. Hobel. 1979. Fetal and meternal copper levels in patients at term with and without premature rupture of membranes. Obstet. Gynecol. 53(5):608-610.

Babu, U. and M.L. Failla. 1990. Respiratory burst and candidacidal activity of peritoneal macrophages are impaired in copper-deficient rats. J. Nutr. 120(12):1692-1699.

Bala, S. and M.L. Failla. 1992. Copper deficiency reversibly impairs DNA synthesis in activated T lymphocytes by limiting interleukin 2 activity. Proc. Natl. Acad. Sci. (USA) 89(15):6794-6797.

Bennetts, H.W., A.B. Beck, and R. Harley. 1948. The pathogenesis of "falling disease": studies of copper deficiency in cattle. Aust. Vet. J. 24(9):237-244.

Brewer, G.J., V.D. Johnson, R.D. Dick, K.J. Fink, K.J. Kluin, and P. Hedera. 2000. Treatment of Wilson's disease with zinc XVII: Treatment during pregnancy. Hepatology 31(2)364-370.

Brown, D., W.J. Schulz-Schaeffer, B. Schmidt, and H.A. Kretzschmar. 1997. Prion protein-deficient cells show altered response to oxidative stress due to decreased SOD-1 activity. Exp. Neurol. 146(1):104-112.

Buamah, P.K., M. Russell, A. Milford-Ward, P. Taylor, and D.F. Roberts. 1984. Serum copper concentrations significantly less in abnormal pregnancies. Clin. Chem. 30(10):1676-1677.

Cohen, N., C.L. Keen, B. Lonnerdal, and L.S. Hurley. 1983. The effect of copper supplementation on the teratogenic effects of triethylenetetramine in rats. Drug Nutr. Interact. 2(3):203-210.

Cordano, A. 1998. Clinical manifestations of nutritional copper deficiency in infants and children. Am. J. Clin. Nutr. 67(5 Suppl.):1012S-1065S.

Cordano, A., J.M. Baertl, and G.G. Graham. 1964. Copper deficiency in infants. Pediatrics 34:324-326.

Danks, D.M. 1988. Copper deficiency in humans. Annu. Rev. Nutr. 8:235-257.

Das, S., B. Levinson, C. Vulpe, S. Whitney, J. Gitschier, and S. Packman S. 1995. Similar splicing mutations of the Menkes/mottled copper-transporting ATPase gene in occipital horn syndrome and the blotchy mouse. Am. J. Hum. Genet. 56(3):570-576.

Dubick, M.A., C.L. Keen, and R.B. Rucker. 1985. Elastin metabolism during perinatal lung development in the copper-deficient rat. Exp. Lung Res. 8(4):227-241.

Dubick, M.A., G.C. Hunger, S.M. Casey, and C.L. Keen. 1987. Aortic ascorbic acid, trace elements, and superoxide dismutase activity in human aneurysmal and occlusive disease. Proc. Soc. Exp. Biol. Med. 184(2):138-143.

Dubick, M.A., C.L. Keen, R.A. DiSilvestro, C.D. Eskelson, J. Ireton, G.C. Hunter. 1999. Antioxidant enzyme activity in human abdominal aortic aneurysmal and occlusive disease. Proc. Soc. Exp. Biol. Med. 220(1): 39-45.

Eum, W.S., and J.H. Kang. 1999. Release of copper ions from the familial amyotrophic lateral sclerosis-associated Cu,Zn-superoxide dismutase mutants. Mol. Cells 9(1):110-114.

Fantel, A.G. 1996. Reactive oxygen species in developmental toxicity: review and hypothesis. Teratology 53(3):196-217.

Gurney, M.E., H. Pu, A.Y. Chiu, M.C. Dal Canto, C.Y. Polchow, D.D. Alexander, J. Caliendo, A. Hentati, Y.W. Kwon, and H.X. Deng. 1994. Motor neuron degeneration in mice tht express a human Cu,Zn superoxide dismutase mutation. Science 264(5166):1772-1775.

Harris, A.L., L.W. Klomp, and J.D. Gitlin. 1998. Aceruloplasminemia: An inherited neurodegenerative disease with impairment of iron homeostasis. Am. J. Clin. Nutr. 67(5 Suppl.):972S-977S.

Harris, Z.L., H. Takahashi, H. Miyajima, M. Serizawa, R.T. MacGillivray, and J.D. Gitlin. 1995. Aceruloplasminemia: molecular characterization of this disorder of iron metabolism. Proc. Natl. Acad. Sci. (USA) 92(7):2539-2543.

Hart, E.B., H. Steenbock, J. Waddell, C.A. Elvehjem. 1928. Iron in nutrition. VII: Copper as a supplement to iron for hemoglobin building in the rat. J. Biol. Chem. 77(April 1):797-812.

Hawk S.N., J.Y. Uriu-Hare, G.P. Daston, and C.L. Keen. 1995. Oxidative damage as a potential mechanism contributing to Cu deficiency-induced defects in rat embryos. Teratology 51(3):171-172.

Howell, J.M., Y. Shunxiang, and J.M. Gawthorne. 1993. Effect of thiomolybdate and ammonium molybdate in pregnant guinea pigs and their offspring. Res. Vet. Sci. 55(2):224-230.

Hunt, C.D., and J.P. Idso. 1995. Moderate copper deprivation during gestation and lactation affects dentate gyrus and hippocampal maturation in immatue male rats. J. Nutr. 125(10):2700-2710.

Hurley, L.S. and C.L. Keen. 1979. Teratogenic effects of copper. Pp. 33-56 in Copper in the Environment. Part II: Health Effects, J.O. Nriagu, ed. New York: John Wiley & Sons.

IOM (Institute of Medicine). 1990. Nutrition during pregnancy. Part I: Weight Gain. Part II: Nutrient Supplements. Washington, DC: National Academy Press.

Jasim, S., B.R. Danielsson, H. Tjalve, and L. Dencker. 1985. Distribution of Cu in foetal and adult tissues in mice: influence of sodium diethyl-

dithiocarbamate. Acta Pharmacol. Toxicol. (Copenhagen) 57(4): 262-270.
Johnson, W.T., and A.C. Thomas. 1999. Copper deprivation potentiates oxidative stress in HL-60 cell mitochondria. Proc. Soc. Exp. Biol. Med.. 221(2):147-152.
Kaler, S.G. 1998. Diagnosis and therapy of Menkes syndrome, a genetic form of copper deficiency. Am. J. Clin. Nutr. 67(5 Suppl.):1029S-1034S.
Kato, T., M. Daimon, T. Kawanami, Y. Ikezawa, H. Sasaki, and K. Maeda. 1997. Islet changes in hereditary ceruloplasmin deficiency. Hum. Pathol. 28(4):499-502.
Keen, C.L. 1996. Teratogenic effects of essential trace metals: deficiencies and excesses. Pp. 977-1001 in Toxicology of Metals, L.W. Chang, L. Magos, and T. Suzuki, eds. New York: CRC Press.
Keen, C.L., N.L. Cohen, B. Lonnerdal, L.S. Hurley. 1983a. Teratogenesis and low copper status resulting from triethylenetetramine in rats. Proc. Soc. Exp. Biol. Med. 173(4):598-605.
Keen, C.L., P. Mark-Savage, B. Lonnerdal, L.S. Hurley. 1983b. Teratogenic effects of D-penicillamine in rats: relation to copper deficiency. Drug Nutr. Interact. 2(1):17-34.
Kelley, D.S., P.A. Daudu, P.C. Taylor, B.E. Mackey, and J.R. Turnlund. 1995. Effects of low-copper diets on human immune response. Am. J. Clin. Nutr. 62(2):412-416.
Kiilholma, P., M. Gronroos, R. Erkkola, P. Pakarinen, and V. Nanto. 1984. The role of calcium, copper, iron and zinc in preterm delivery and premature rupture of fetal membranes. Gynecol. Obstet. Invest. 17(4):194-201.
Klevay, L.M. 1998. Lack of a recommended dietary allowance for copper amy be hazardous to your health. J. Am. Coll. Nutr. 17(4):322-326.
Kondo, T., A.G. Reaume, T.T. Huang, E. Carlson, K. Murakami, S.F. Chen, E.K. Hoffman, R.W. Scott, C.J. Epstein, and P.H. Chan. 1997. Reduction of CuZn-superoxide dismutase activity exascerbates neuronal cell injury and edem formation after transient focal cerebral ischemia. J. Neurosci. 17(11):4180-4189.
Kuznetsov, A.V., J.F. Clark, K. Winkler, W.S. Kunz. 1996. Increase of flux control cytochrome c oxidase in copper-deficient mottled brindled mice. J. Biol. Chem. 271(1):283-288.
Lane, T.F., M.L. Iruela-Arispe, R.S. Johnson, and H. Sage. 1994. SPARC is a source of copper-binding peptides that stimulate angiogenesis. J. Cell Biol. 125(4):929-943.
Lee S.H., R. Lancey, A. Montaser, N. Madani, M.C. Linder. 1993. Ceruloplasmin and copper transport during the latter part of gestation in the rat. Proc. Soc. Exp. Biol. Med. 203(4):428-39.
Mark-Savage, P., C.L. Keen, and L.S. Hurley. 1983. Reduction by copper

supplementation of teratogenic effects of D-penacillamine. J. Nutr. 113(3):501-510.
Medeiros, D.M., R.E. Wildman. 1997. Newer findings on a unified perspective of copper restriction and cardiomyopathy. Proc. Soc. Exp. Biol. Med. 215(4):299-313.
Melov, S., J.A. Schneider, B.J. Day, D. Hinerfeld, P. Coskun, S.S. Mirra, J.D. Crapo, and D.C. Wallace. 1998. A novel neurological phenotype in mice lacking mitochondrial manganase superoxide dismutase. Nat. Genet. 18(2):159-163.
Mercer, J.F. 1998. Menkes syndrome and animal models. Am. J. Clin. Nutr. 67(5 Suppl.):1022S-1028S.
Mercer, J.F., A. Grimes, L. Ambrosini, P. Lockhart, J.A. Paynter, H. Dierick, and T.W. Glover. 1994. Mutations in murine homologue of the Menkes gene dappled and blotchy mice. Nat. Genet. 6(4):374-378.
Mills, C.F., and R.B. Williams. 1962. Copper concentrations and cytochrome-oxidase and ribonuclease activities in the brains of copper deficient lambs. Biochem. J. 85:629-632.
Milne, D.B. 1998. Copper intake and assessment of copper status. Am. J. Clin. Nutr. 67(5 Suppl.):1041S-1045S.
Milne, D.B., and F.H. Nielsen. 1996. Effects of a diet low in copper on copper status indicators in postmenopausal women. Am. J. Clin. Nutr. 63(3):358-364.
Miyajima, H., Y. Takahashi, M. Serizawa, E. Kaneko, and J. D. Gitlin. 1996. Increased plasma lipid peroxidation in patients with aceruloplasminemia. Free Radic. Biol. Med. 20(5):757-760.
Morton, M.S., P.C. Elwood, and M. Abernethy. 1976. Trace elements in water and congenital malformations of the central nervous system in South Wales. Br. J. Prev. Soc. Med. 30(1):36-39.
Neal, W.M., R.B. Becker, and A.L. Shealy. 1931. A natural copper deficiency in cattle rations. Science 74(1921):418-419.
NRC (National Research Council). 1989. Recommended Dietary Allowances, 10th Ed. Washington, DC: National Academy Press.
Olivares, M., and R. Uauy. 1996. Copper as an essential nutrient. Am. J. Clin. Nutr. 63(5 Suppl.):791S-796S.
Pauly, P.C., and D.A. Harris. 1998. Copper stimulates endocytosis of the prion protein. J. Biol. Chem. 273(50):33107-33110.
Percival, S.S. 1998. Copper and immunity. Am. J. Clin. Nutr. 67(5 Suppl.):1064S-1068S.
Prohaska, J.R., and W.R. Bailey. 1995. Alterations of rat brain peptidylglycine a-amidating monooxygenase and other cuproenzyme activities following perinatal copper deficiency. Proc. Soc. Exp. Biol. Med. 210(2):107-116.
Prohaska, J.R., and R.G. Hoffman. 1996. Auditory startle response is di-

minished in rats after recovery from perinatal copper deficiency. J. Nutr. 126(3):618-627.

Prohaska, J.R., and W.W. Wells. 1975. Copper deficiency in the developing rat brain: Evidence for abnormal mitochondria. J. Neurochem. 25(3):221-228.

Prohaska, J.R., T. Tamura, A.K. Percy, and J.R. Turnlund. 1997. In vitro copper stimulation of plasma peptidylglycine a-amidating monooxygenase in Menkes disease variant with occipital horns. Pediatr. Res. 42(6):862-865.

Proud, V.K., H.G. Mussell, SG. Kaler, D.W. Young, and A.K. Percy. 1996. Distinctive Menkes disease variant with occipital horns: Delineation of natural history and clinical phenotype. Am. J. Med. Genet. 65(1):44-51.

Reiser, S., J.C.J. Smith, W. Mertz, J.T. Holbrook, D.J. Scholfield, A.S. Powell, W.K. Canfield, and J.J. Canary. 1985. Indices of copper status in humans consuming a typical American diet containing either fructose or starch. Am. J. Clin. Nutr. 42(2):245-251.

Romero, N., D. Tinker, D. Hyde, and R.B. Rucker. 1989. Role of plasma and serum proteinases in the degradation of elastin. Arch. Biochem. Biophys. 244(1):161-168.

Rosa, F.W. 1986. Teratogen update: Penacillamine. Teratology 33(1):127-31.

Rossi, L., G. Lippe, E. Marchese, A. De Martino, I. Mavelli, G. Rotilio, M.R. Ciriolo. 1998. Decrease in cytochrome c oxidase protein in heart mitochondria of copper-deficient rats. Biometals 11(3):207-212.

Rucker, R.B., T. Kosonen, M.S. Clegg, A.E. Mitchell, B.R. Rucker, J.Y. Uriu-Hare, and C.L. Keen. 1998. Copper, lysyl oxidase, and extracellular matrix protein cross-linking. Am. J. Clin. Nutr. 67(5 Suppl.): 996S-1002S.

Salgo, M.P., and G. Oster. 1974. Fetal resorption induced by disulfiram in rats. J. Reprod. Fertil. 39(2):375-377.

Shavlovski, M.M., N.A. Chebotar, L.A. Konopistseva, E.T. Zakharova, A.M. Kachourin, V.B. Vassiliev, and V.S. Gaitskhoki. 1995. Embryotoxicity of silver ions is diminished by ceruloplasmin—Further evidence for its role in the transport of copper. Biometals 8(2):122-128.

Shaw, J.C.L. 1992. Copper deficiency in term and preterm infants. Pp. 105-119 in Nestle Nutrition Workshop Series. Vol. 30, Nutritional Anemias, S.J. Fomon, and S. Zlotkin, eds. New York: Raven Press.

Speich, M., A. Murat, J.L. Auget, B. Bousquet, and P. Arnaud. 1992. Magnesium, total calcium, phosphorous, copper and Zn in plasma and erythrocytes of venous cord blood from infants of diabetic mothers: Comparison with a reference group by logistic discriminant analysis. Clin Chem. 38(10):2002-2007.

Takahashi, Y., H. Miyajima, S. Shirabe, S. Nagataki, A. Suenaga, and J.D.

Gitlin. 1996. Characterization of a nonsense mutation in the ceruloplasmin gene resulting in diabetes and neurodegeneragtive disease. Hum. Mol. Genet. 5(1):81-84.

Taubeneck, M.W., J.L. Domingo, J.M. Llobet, and C.L. Keen. 1992. Meso-2,3-dimercaptosuccinic acid (DMSA) affects maternal and fetal copper metabolism in Swiss mice. Toxicology 72(1):27-40.

Tokuda, Y., S. Yokoyama, M. Tsuji, T. Sugita, T. Tajima, and T. Mitomi. 1986. Copper deficiency in an infant on prolonged total parenteral nutrition. J. Parenter. Enteral. Nutr. 10(2):242-244.

Troy, C.M., and M.L. Shelanski. 1994. Down-regulation of copper/zinc superoxide dismutase causes apoptotic death in PC12 neuronal cells. Proc. Natl. Acad. Sci. (USA) 91(14):6384-6387.

Turnlund, J.P. 1994. Copper. Pp. 231-241 in Modern Nutrition in Health and Disease, Vol. I., 8th Ed., M.E. Shils, J.A. Olson, and M. Shike, eds. Philadelphia: Lea and Febiger.

Uauy, R., M. Olivares, and M. Gonzalez. 1998. Essentiality of copper in humans. Am. J. Clin. Nutr. 67(5 Suppl.):952S-959S.

Uriu-Hare, J.Y., J.S. Stern, and C.L. Keen. 1989. Influence of maternal dietary Zn intake on expression of diabetes-induced teratogenicity in rats. Diabetes 38(10):1282-1290.

Weisenberg, E., A. Harbreich, and J. Mager. 1980. Biochemical lesions in copper-deficient rats caused by secondary iron deficiency. Derangement of protein synthesis and impairment of energy metabolism. Biochem. J. 188(3):633-641.

Wiener, G., I. Wilmut, and A.C. Field. 1978. Maternal and lamb breed interactions in the concentration of copper in tissues and plasma of sheep. Pp. 469-472 in Trace Element Metabolism in Man and Animals—3, M. Kirchgessner, ed. Freising-Weihenstephan, Germany: Technische Universitat Munchen.

Yoshida, K., K. Furihata, S. Takeda, A. Nakamura, K. Yamamoto, H. Morita, S. Hiyamuta, S. Ikeda, N. Shimizu, and N. Yanagisawa. 1995. A mutation in the ceruloplasmin gene is associated with systemic hemosiderosis in humans. Nat. Genet. 9(3):267-272.

Yoshikawa, H., H. Seo, T. Oyamada, T. Ogasawara, T. Oyamada, T. Yoshikawa, X. Wei, S. Wang, and Y. Li. 1996. Histopathology of enzootic ataxia in Sika deer (Cervus nippon Temminck). J. Vet. Med. Sci. 58:849-854.

Zidenberg-Cherr, S., and C.L. Keen. 1991. Essential trace elements in antioxidant processes. Pp. 107-127 in Trace Elements, Micronutrients and Free Radicals, I.E. Dreosti, ed. Clifton, NJ: Human Press.

Zidenberg-Cherr, S., P.A. Benak, L.S. Hurley, and C.L. Keen. 1988. Altered mineral metabolism: A mechanism underlying the fetal alcohol syndrome in rats. Drug Nutr. Interact. 5(4):257-274.

4

Disorders of Copper Homeostasis

DISORDERS in copper homeostasis are discussed in this chapter. The two best-studied disorders in copper regulation, Menkes disease and Wilson disease, are described first, including the current state of knowledge of the genetics that underlie these two disorders. Occipital horn syndrome, a milder form of Menkes disease, is then discussed. Aceruloplasminemia, a newly recognized disease caused by a defect in the gene encoding ceruloplasmin, is then discussed. Other syndromes described are Tyrolean infantile cirrhosis (TIC), Indian childhood cirrhosis (ICC), and idiopathic copper toxicosis (ICT). All are associated with ingestion of high amounts of copper and might have a genetic component. Finally, disease-induced changes in copper homeostasis are discussed.

MENKES DISEASE

Menkes disease was originally diagnosed as an X-linked neurodegenerative disorder in infants characterized by poor growth and unusual "kinky" hair texture (Menkes et al. 1962). The defect is manifested primarily in males because it is X linked; eight females, however, have also been reported to have the disease (Kodama and Murata 1999). The major defect in Menkes disease is a failure to transport copper ions completely across the intestinal mucosa, ultimately leading to a severe copper deficiency in the peripheral organs (Danks et al. 1972). Transport across the blood-brain barrier is also impaired.

Clinically, Menkes patients usually have low plasma ceruloplasmin concentrations and decreased concentrations of copper in the liver and brain (Danks 1995). The morphological changes of the disease start in utero and are fully manifested during the perinatal period (Vulpe and Packman 1995). Premature birth and being small for gestational age are frequent characteristics of Menkes patients. Hypothermia, prolonged jaundice, feeding difficulties, and diarrhea can occur in the neonatal period (Horn et al. 1992). Developmental delays are apparent around the third month, as shown by abnormal head movement and the absence of a smiling response. Therapy-resistant convulsions also occur (Horn et al. 1992). There is no evidence that maternal copper status influences the development or expression of Menkes disease in the infant.

The primary genetic defect in Menkes disease is in the protein ATP7A, a membrane-bound Cu-ATPase that regulates the outward flow of copper ions from the interior to the exterior of the cell (Chelly et al. 1993; Mercer et al. 1993; Vulpe et al. 1993). For this reason, Menkes patients accumulated copper in the intestinal cells. Skin fibroblasts from patients with Menkes disease accumulate large amounts of copper when grown for 3 to 5 days in essential growth medium, demonstrating that the primary defect is detectable at the cellular level (Goka et al. 1976; Horn 1976). Thus, copper accumulation by fibroblasts has been used to confirm prenatal and postnatal diagnosis of Menkes disease. Unlike normal fibroblasts, fibroblasts from Menkes patients became growth-sensitive in very-low-to-medium copper concentrations during a 7-day exposure (Rayner and Suzuki 1994). No significant differences are detectable in copper accumulation by cells established from Menkes patients with different allelic variants of the disease. Thus, the severity of the phenotype cannot be determined by in vitro copper uptake assays (Masson et al. 1997).

In the case of mucosal epithelia, the ATPase transports the copper ions into the serosal capillaries as part of the absorption mechanism across the gut. Studies with animal models and isolated cell cultures suggest that the neurodegeneration is caused by a failure of the ATPase that normally transports copper, resulting in reductions in the activity of select copper-dependent enzymes (Kodama 1993; Yoshimura et al. 1995; Qian et al. 1997; Qian et al. 1998).

Efforts to treat Menkes disease by giving the patients parenteral copper in the form of copper histidine, copper acetate, or copper EDTA have had little success to date. None of these agents prevents neurological damage, although they can increase serum copper concentrations and ceruloplasmin activity. The age when copper administration begins seems to be of some consequence. In one study, a Menkes patient with a splice acceptor site mutation in a non critical region was treated with copper histidine beginning on day 8 of life. Head growth, myelination of neurons in the brain, and neurodevelopment were all normal in the patient. A half brother and a cousin with the same mutation showed arrested head growth, cerebral atrophy, delayed myelination, and abnormal neurodevelopment (Kaler et al. 1996). Injections of copper histidine after 1-month of age failed to arrest the neurodegeneration in another Menkes patient (Sarkar et al. 1993). If treated early, some Menkes patients (those with only a partial gene knockout) given copper histidine survive beyond their teens (Christodoulou et al. 1998). To be effective, copper histidine must be administered at a dose of 200-1,000 µg per day, once per day or 2-3 times per week (Kaler et al. 1996). The adequacy of the dose is determined by measuring urinary copper excretion, serum copper concentrations and ceruloplasmin activity, as well as copper concentrations in the liver. An alternative and potentially effective therapy uses lipid-soluble chelators of copper, such as diethyldithiocarbamate or dimethyldithiocarbamate, administered intraperitoneally even without copper. That treatment increased the survival time of macular mutant mice, presumably by enhancing copper transport across cellular membranes (Tanaka et al. 1990). There is no record of the use of these chelators to treat Menkes patients.

The X-linked mottled locus in mice ($Atp7a^{Mo}$) is the homolog of the Menkes gene (Cecchi and Avner 1996). Mottled mutants exhibit a phenotype that closely resembles that seen in Menkes patients. The mottled locus spans 120 kilobase (kb) and encodes a gene for a copper-transporting ATPase that is 89.9% identical to $ATP7A$ (Mercer et al. 1993). Spontaneous mutations at the mottled locus cause reduced fertility and viability and lead to phenotypic symptoms of classical Menkes disease and occipital horn syndrome (OHS). Of the 20 independent mottled alleles that have been identified, 10 are spontaneous and 14 arose after gamma or X irradiation (Cecchi and Avner 1996). The six allelic variants appearing most often in the literature are brindled ($Atp7a^{Mo-br}$), macular ($Atp7a^{Mo-ml}$), blotchy ($Atp7a^{Mo-bo}$), viable brindled ($Atp7a^{Mo-vbr}$), tortoiseshell ($Atp7a^{Mo-to}$), and dappled ($Atp7a^{Mo-dp}$). Males carrying the $Atp7a^{Mo-ml}$ and $Atp7a^{Mo-br}$ survive only a few days after birth, and males carrying $Atp7a^{Mo-vbr}$ and $Atp7a^{Mo-blo}$ survive for several months. Mice hemizygous for the blotchy allele have connective-tissue defects and only 30% of the normal lysyl oxidase activity (Royce et al. 1982). The mottled mutant mouse, therefore, is considered an animal model for occipital horn syndrome (OHS).

OCCIPITAL HORN SYNDROME

Mutations in the Menkes gene can also result in OHS, formerly called X-linked cutis laxia. OHS is a milder form of classical Menkes disease in which the same copper transporter, ATP7A, is affected. Because it affects the same gene locus (Xq13.3), OHS is referred to as an allelic variant of

Menkes disease (Kodama and Murata 1999). Like Menkes, OHS is recessively inherited and characterized by abnormalities in copper metabolism (Proud et al. 1996). Whereas the prevalence rate of Menkes disease appears to be 1 in 100,000-250,000 births, the prevalence rate of OHS appears to be much less. For example, only a few cases have been reported in Japan (Kodama et al. 1999).

Although Menkes and OHS have a similar abnormality in copper metabolism, the two diseases have different clinical presentations and survival potentials (Das et al. 1995). Because of late and inconsistent onset of symptoms, OHS might escape early recognition. A patient must be at least 2 years of age before a definitive diagnosis of OHS is possible (De Paepe et al. 1999). A prominent protuberance (exostoses) in the occipital bone seen on radiographs defines the condition (Proud et al. 1996). That protuberance is not seen in patients with classical Menkes disease. Additional skeletal abnormalities are short and broad clavicles, osteoporosis, laxity of the skin and joints, and bladder diverticula (Kodama et al. 1999). Serum copper and ceruloplasmin are low in some but not all patients, and cultured skin fibroblasts typically have low lysyl oxidase activity. Although disturbance in electroencephalogram patterns are seldom seen, arrested mental development and late-onset seizures often accompany the disease. Most OHS patients suffer inguinal hernias, recurrent diarrhea, urinary-tract infections, and distorted facial features, such as high foreheads and hooked noses. Radiographs show tortuous cerebral blood vessels with multiple branch occlusions (Proud et al. 1996). Although OHS patients can survive until adulthood, they have borderline intelligence. To date, no published reports have appeared describing parenteral administration of copper to correct, manage, or improve OHS patients (Kodama et al. 1999).

Studies of cells taken from OHS patients have linked *MNK* gene processing with the development of OHS. The human *MNK* gene normally has three 98-base pair (bp) tandem repeats in an upstream promoter region (Levinson et al. 1996). Genomic DNA from an OHS patient was found to have only 2 of the 98-bp repeats; no other mutations were found. The 98-bp deletion led to a dramatic decrease in the expression of a CAT reporter gene construct, suggesting that the 98-bp repeats play a role in regulating *MNK* mRNA expression (Levinson et al. 1996). Northern blot analysis, however, revealed no detectable reduction in *MNK* mRNA levels in cells derived from the patient. In another patient with OHS, an A-T transversion in intron 10 resulted in a loss of exon 10 by alternative splicing of *ATP7A* mRNA (Qi and Byers 1998). The mutated gene gave rise to an *ATP7A* variant that lacked two of the eight transmembrane domains and, for reasons not understood, remained resident in the endoplasmic reticulum and was not transported to the Golgi. To be effective in export, the Cu-ATPase must enter into the secretory pathway by first moving from the endoplasmic reticulum to the Golgi (Francis et al. 1998). Because the error was not observed in all the mRNAs, the individual had the capacity to synthesize an intact Cu-ATPase, but at a reduced level. The observation might explain why OHS is a considered a mild as opposed to a debilitating form of Menkes disease.

Tissues from mottled mutants bearing the blotchy allele display two large sized mRNAs, demonstrating a likely defect in splicing (Mercer et al. 1993). Levinson et al. (1996) showed that the splice-site mutation in the blotchy allele is identical to that seen in mRNA from Menkes patients.

WILSON DISEASE

Wilson disease, or hepatolenticular degeneration, is an autosomal recessive disorder that results from accumulation of copper predominantly in the liver and brain (Bush et al. 1955; Strickland et al. 1969; O'Reilly et al. 1971; Frommer 1974; Gibbs and Walshe 1980). The accumulation is due to defective biliary excretion of copper. Current data indicate that adult humans need to ingest about 0.75 mg of copper daily to sustain a balance (Brewer and Yuzbasiyan-Gurkan 1992). Typically, humans ingest about 1 mg of copper per day (Holden et al. 1979; Klevay et al. 1979; Reiser et al. 1985; Hill et al. 1987; Iyengar et al. 1988; Milne 1998; Brewer and Yuzbasiyan-Gurkan 1992). The daily excess of copper averaging about 0.25 mg per day is normally excreted in the feces (Brewer and Yuzbasiyan-Gurkan 1992). However, due to a genetic defect, individuals with Wilson disease are unable to excrete the excess copper, resulting in a gradual build-up of copper in the body (Bush et al. 1955; Frommer 1974; O'Reilly et al. 1971; Strickland et al. 1969; Gibbs and Walshe 1980).

Clinically, Wilson patients begin receiving medical attention because of hepatic, neurological, or psychiatric symptoms, in roughly equal proportions (Brewer and Yuzbasiyan-Gurkan 1992; Scheinberg and Sternlieb 1984). Some individuals with Wilson disease are diagnosed before the appearance of clinical symptoms through the screening of Wilson disease siblings, who are at a 25% risk for the disease (Brewer and Yuzbasiyan-Gurkan 1992; Scheinberg and Sternlieb 1984). Wilson patients have underlying liver cirrhosis at the time of diagnosis, even if the symptoms are subclinical (Brewer and Yuzbasiyan-Gurkan 1992).

The primary genetic defect in Wilson disease is in *ATP7B*, which encodes a copper transport protein (Bull et al. 1993; Tanzi et al. 1993; Yamaguchi et al. 1993). The normal mechanism for biliary excretion of copper and the role of the *ATP7B* gene product in that excretion have yet to be elucidated. It has been hypothesized that ceruloplasmin plays a role in copper excretion by preventing the reabsorption of copper (Iyengar et al. 1988). That hypothesis is supported by evidence of a high-molecular-weight protein in the bile that reacts with antibodies to ceruloplasmin, and contains enough copper to account for the excess copper (Iyengar et al. 1988; Chowrimootoo et al. 1996; Davis et al. 1996). There is no evidence of immunoreactive ceruloplasmin in the bile of most Wilson patients (Iyengar et al.1988), and Wilson patients usually have low blood ceruloplasmin levels (Brewer and Yuzbasiyan-Gurkan 1992). It is not known, however, how the genetic defect in *ATP7B* and the effects on ceruloplasmin in Wilson patients are related. It is important to note that patients with aceruloplasminemia have normal copper concentrations (Harris et al. 1998).

Although diets low in copper have been prescribed for patients with Wilson disease in the past, more recent measurements of food copper content have indicated that only liver and shellfish are high enough in copper to warrant restriction (Brewer and Yuzbasiyan-Gurkan 1992). If the concentration of copper in drinking water at home, school, or work is greater than 0.1 mg per liter (L), most clinicians recommend that alternate sources be used in the management of Wilson disease (Brewer et al. 1998).

GENETIC CHARACTERISTICS OF WILSON AND MENKES DISEASES

The genetic mutations responsible for the loss of copper homeostasis are well characterized for Menkes and Wilson diseases. As a result, both diseases have provided insight into the genetic factors that regulate copper bioavailability and transport to organs and tissues.

The identification, sequencing, cloning, and characterization of the Wilson gene and mutations of the gene (Bull et al. 1993; Tanzi et al. 1993; Petrukhin et al. 1993; Thomas et al. 1995) and the Menkes gene (Vulpe et al. 1993; Chelly et al. 1993; Mercer et al. 1993) reveal that both genes encode P-type Cu-ATPases that are specific for copper transport (Solioz and Vulpe 1996). Cu-ATPases are complex integral membrane proteins that are part of a family of ion-transporting proteins that include the Ca^{2+} and Na^+/K^+ transport proteins. The Wilson protein (ATP7B) and the Menkes protein (ATP7A) share 57% amino acid sequence homology (Vulpe and Packman 1995) and show remarkable similarity to bacteria copper-binding proteins (Silver et al. 1993).

The Menkes gene, *MNK*, spans about 140-150 kb (Tumer et al. 1995; Dierick et al. 1995) and gives rise to a 8.5-kb mRNA. The Menkes mRNA encodes a protein (ATP7A) of exactly 1,500 amino acids, but additional nucleotide sequences at the 5' end are also known to occur (Tumer et al. 1995). High levels of *MNK* mRNA are found in muscle, kidney, lung, and brain, and low levels are found in the placenta and pancreas, and only trace amounts are found in the liver (Chelly et al. 1993; Mercer et al. 1993). A number of mutations can result in Menkes disease. For example, out of more than 300 Menkes cases from 27 different countries, 191 unique mutations of the Menkes gene (*ATP7A*) were found (Tumer et al. 1995). Thirty-five of the mutations were partial gene deletions and 149 were point mutations. The largest gene deletion involved the whole gene except for the first two exons; the smallest was a deletion of exon 1. The four types of point mutations—namely, deletion/insertion, missense, nonsense, and splice site—were represented almost equally.

The Wilson transcript is 7.5 kb and encodes a protein of 1,411 amino acids. The Wilson gene is strongly expressed in the kidney, but unlike the Menkes gene, it is also strongly expressed in liver (Bull et al. 1993; Yamaguchi et al. 1993). The chromosomal locus of the Wilson gene is 13q14.3-q21. More than 40 normal allelic variants and 170 disease-associated variants have been reported for the Wilson gene (Cox and Roberts 1999).

The structure and function of the two gene products are similar. Both genes affect copper trafficking, but at different sites and in opposing directions. Immunochemical studies localized ATP7A to the perinuclear area thought to represent the Golgi (Dierick et al. 1997). From there, the Menkes protein regulates the release of copper ions into extracellular spaces by actively transporting the copper into export vesicles (Voskoboinik et al. 1998). Overexpression of the ATPase allows cells to tolerate higher than normal amounts of copper in their environment (Camakaris et al. 1995). The vesicles are not fixed but relocate to the plasma membrane in response to high extracellular copper. Some investigators postulate that ATP7A-laden vesicles continually move between the Golgi and the plasma membrane (Camakaris et al. 1995). The movement is a function of the primary structure of ATP7A. Specifically, the GMTCXXC motifs in the heavy metal binding region are important in sensing copper and triggering vesicle movement (Voskoboinik et al. 1999). In experiments using fluorescent antibodies against ATP7A, agents that disrupt Golgi structure or interfere with transport functions were observed to cause a more-dispersed pattern of fluorescence, indicating interference with the localization or aggregation of ATP7A within the secretory vesicles.

The Wilson protein (ATP7B) appears to be located on the Golgi and mitochondrial membrane (Lutsenko and Cooper 1998). In contrast to the Menkes protein, ATP7B is involved in the intracellular transport of copper (Terada and Sugiyama 1999). Given the above, it is evident that dysfunction of either ATP7A or ATP7B results in transport impairment and a disruption of copper distribution.

Regulation of Wilson and Menkes gene expression has not received much attention. Studies of a 1.3-kb segment in the 5'-flanking region of the Wilson gene found evidence of four metal response elements (MREs) and six MRE-like sequences (MLSs) that bear structural resemblance to elements found in the metallothionein gene (Oh et al. 1999). There is evidence for regulatory elements such as Sp1, AP-1, AP-2, and E-box. The data suggest that the Wilson gene contains a single transcription site and has strong cis-acting elements at −811 to −653 from the transcription initiation site for high expression of the Wilson gene (Oh et al. 1999).

With a frequency of 1 per 7,000 live births, Sardinia has the highest reported incidence of Wilson disease of any population of people of Mediterranean descent (Loudianos et al. 1999). Mutation analysis in that population led to the characterization of 13 rare and 2 common mutations, together accounting for about 30% of the chromosome. The most common haplotype, however, occurs in 60.5% of the chromosome. Recently, it was learned that that mutation is not in the coded part of the gene but in regulatory elements in the 5' UTR (Loudianos et al. 1999). Sequencing the promoter GC-rich region in 92 chromosomes with the haplotype led to the discovery of a single mutation manifested as a 15-nucleotide deletion from position −441 to position −427 relative to the translation start site. A construct containing the deletion was only 25% as effective as a control in a luciferase reporter activity assay (Loudianos et al. 1999). Thus, in the Sardinian population, Wilson disease in people displaying the most common haplotype was initiated by an apparent failure to express an intact ATP7B protein at levels sufficient to maintain normal copper homeostasis.

HETEROZYGOTES FOR WILSON DISEASE

Because Wilson disease is an autosomal recessive disorder, it is exhibited only in individuals who are homozygous or compound heterozygotes[1] for the gene defect. However, if the phenotype due to defects in the Wilson gene is defined as abnormal copper metabolism, then the Wilson gene can be considered a codominant gene, as abnormalities in copper homeostasis often occur in heterozygous carriers (Yuzbasiyan-Gurkan et al. 1991; Brewer and Yuzbasiyan-Gurkan 1992). Urinary copper concentrations are increased in about half of the calculated number of heterozygous siblings

---

[1]Individuals who have two different mutations of the same gene. One mutation is on one of the paired chromosomes, and the second is on the other chromosome.

of patients with Wilson disease.[2] The siblings' concentrations approached the diagnostic concentration for Wilson disease (100 μg/24 hr) (see Figure 4-1) (Yuzbasiyan-Gurkan et al. 1991; Brewer and Yuzbasiyan-Gurkan 1992). Increased hepatic copper concentrations, which again approach the diagnostic concentration for Wilson disease, in siblings of Wilson patients (200 μg/g of dry weight) (see Figure 4-2) are likely due to heterozygosity for the Wilson gene (Brewer and Yuzbasiyan-Gurkan 1992). In addition, the rate of incorporation of ^{64}Cu into ceruloplasmin in heterozygous individuals is, on average, about half that in normal individuals but can be close to the rate seen in patients with Wilson disease in some heterozygous individuals (Brewer and Yuzbasiyan-Gurkan 1992) (see Figure 4-3). The data in Figure 4-3 indicate that, with typical dietary copper intake, heterozygous carriers can accumulate copper at concentrations only slightly below those characteristic of Wilson disease. Given those data, it is possible that Wilson disease in infants could act as an autosomal dominant disease in cases where copper exposure is high (Brewer in press) (see Sensitive Populations section in Chapter 5 for a further discussion of that hypothesis).

Wilson disease has a reported incidence of about 1 in 40,000 births (Bachmann et al. 1979; Scheinberg and Sternlieb 1984; Giagheddu et al. 1985; Reilly et al. 1993). Using the Hardy-Weinberg equation ($p^2 + 2pq + q^2 = 1$), where p^2 is the frequency of normal homozygotes, $2pq$ is the frequency of heterozygotes for the mutation, and q^2 is the frequency of homozygotes for the mutation and setting q^2 at 1 in 40,000, individuals who are heterozygous for Wilson disease can be calculated to make up 1% of the general population.

The above calculated frequency of Wilson-disease heterozygotes is based on the estimated incidence from Wilson-disease studies cited above. However, Wilson disease is probably underdiagnosed because of the lack of adequate screening for the disease. Currently, about 20% of chronic cirrhosis cases are idiopathic (often called "cryptogenic"), and it is possible that Wilson disease remains undiagnosed among those individuals. In addition, liver disease in some patients who are thought to have hepatitis C might actually be due to undiagnosed Wilson disease. If a large number of patients with liver disease have undiagnosed Wilson disease, the heterozygote frequency might be as high as 2%.

---

[2]Specific individuals were not genotyped as heterozygous.

60

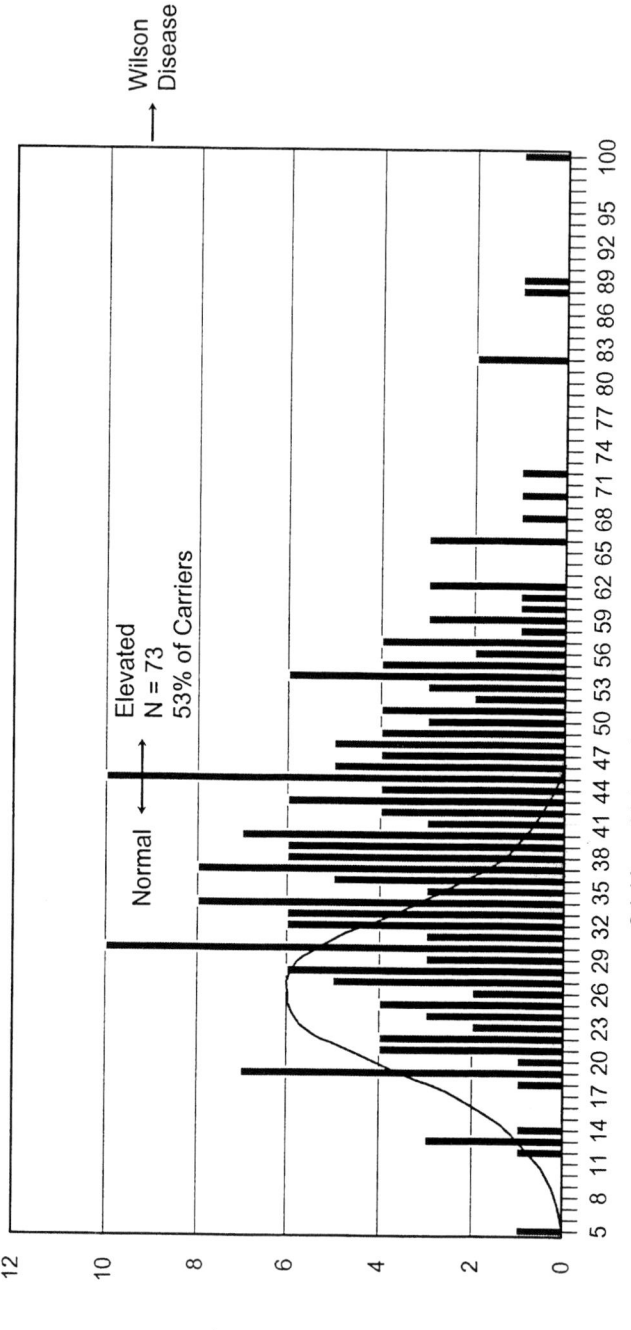

FIGURE 4-1 Frequency distribution of 24-hr urinary copper concentrations for 206 non-Wilson-disease siblings of Wilson-disease patients. This histogram shows the distribution of urinary copper values in 206 siblings of diagnosed Wilson disease patients. The siblings are not phenotypically homozygous for Wilson disease. Probabilistically, 2/3 (138) of the siblings should be carriers of the Wilson-disease gene. Of the estimated 138 carriers, 73 (53%) have increased urinary copper concentrations. The theoretical curve for urinary copper concentrations in normal individuals is superimposed on the left side of the distribution. Source: Adapted from Brewer, in press.

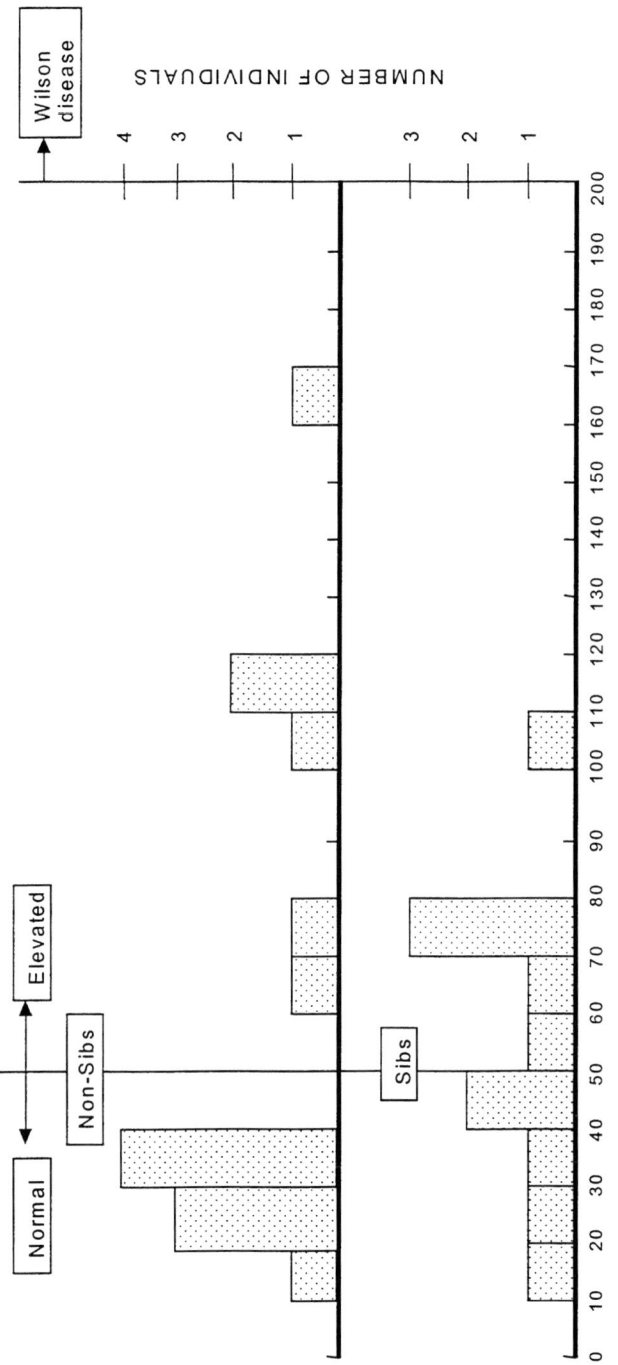

FIGURE 4-2 Hepatic copper concentrations in individuals presumed to be heterozygous for the Wilson-disease gene. The vertical line at 50 μg indicates the upper limit of the normal copper concentration range. The vertical line at 200 μg indicates the lower limit of the copper concentration at which a diagnosis of Wilson disease can be made. The lower panel shows the liver copper concentrations in the siblings (n = 11) of Wilson patients, two-thirds of whom would be expected to be heterozygotes on a probabilistic basis (with 11 individuals, 7 or 8 heterozygotes would be expected). The copper concentration is above the normal range in 6 of the individuals. The upper panel shows the liver copper concentrations in other individuals (n = 14) suspected of having Wilson disease. None of those 14 persons had Wilson disease, but hepatic copper concentrations were increased in 6 individuals. Those 6 individuals are likely heterozygous for the Wilson-gene defect. Source: Adapted from Brewer, in press.

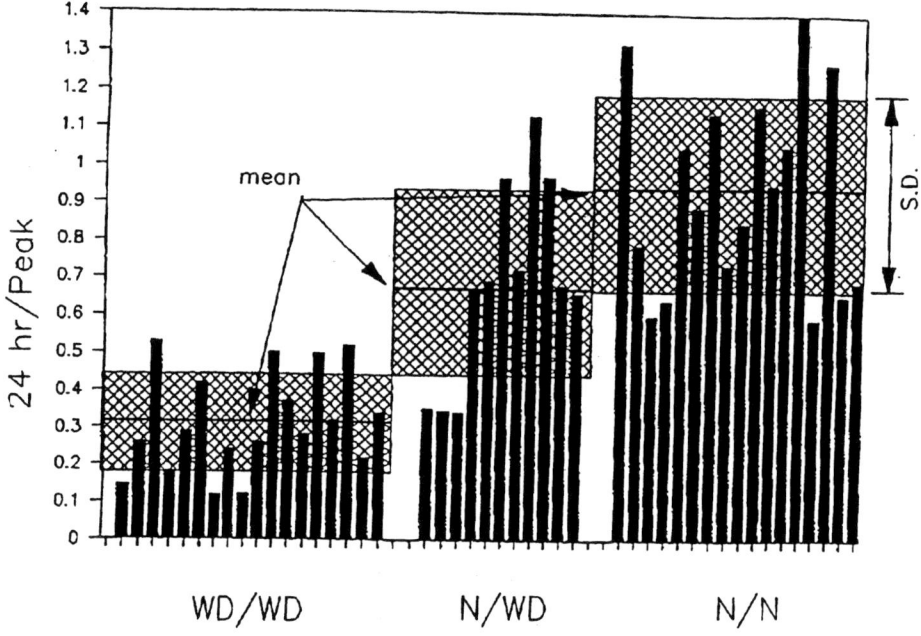

FIGURE 4-3 Incorporation of orally administered ^{64}Cu into ceruloplasmin at 24 hr. Means and standard deviations are shown. The y axis shows the ratio of the 24-hr incorporation of radiocopper over the peak incorporation of radiocopper (at 1 or 2 hr). WD/WD, homozygous affected; N/WD, heterozygous; N/N, homozygous normal. Hatched portion = mean ± standard deviation. Source: Brewer and Yuzbasiyan-Gurkan 1992. Reprinted with permission from *Medicine*; copyright 1992, Lippincott Williams & Wilkins.

ACERULOPLASMINEMIA

Aceruloplasminemia is an autosomal recessive disorder of iron metabolism characterized by a defect in the gene coding for ceruloplasmin. This disease is rare; a frequency of only 1 per 2,000,000 in cases involving non-consanguineous marriages was reported in Japan (Miyajima et al. 1999). Aceruloplasminemic individuals have no oxidase-detectable or immunoreactive ceruloplasmin in their serum (Miyajima et al. 1987). Late-onset retinal and basal ganglia degeneration, diabetes, and neurological symptoms are commonly seen in clinics. The pathogenesis of the disease has been linked to a slow accumulation of iron in tissues (Yazaki et al. 1998; Gitlin 1998). Biopsy examinations have detected unusually high amounts of iron in the pancreas, heart, kidney, spleen, and thyroid gland (Yoshida et al. 1995), and magnetic resonance imaging of the brain shows an increased iron content of the basal ganglia, thalamus, and dentate nucleus

(Yazaki et al. 1998). Tissue copper concentrations, however, tend to be unchanged and total iron-binding capacity and erythrocyte counts are normal. ^{59}Fe administered as a trace dose tends to accumulate in the brain, heart, kidney, and liver of patients with aceruloplasminemia, confirming the biopsy reports. ^{64}Cu administered intravenously, however, shows no such tendency, and tissue copper remains basically normal (Logan et al. 1994). Intravenous injections of human ceruloplasmin raise the serum iron in these patients (Logan et al. 1994). The lack of cardinal copper-related symptoms in aceruloplasminemic individuals challenges what has been considered the essential role of ceruloplasmin in copper transport and homeostasis and has increased the focus on its role.

Despite similarities with hemosiderosis and other iron-overload disorders, aceruloplasminemia is unusual in showing major alterations in neurological functions. Patients with aceruloplasminemia eventually succumb to the effects of increased iron in the tissues, particularly the basal ganglia (Miyajima et al. 1998). An increased susceptibility to lipid peroxidation is believed to contribute substantially to the neuropathology, which suggests that free-radical-mediated tissue injury is responsible for the basal-ganglia degeneration (Miyajima et al. 1996, 1998). Ceruloplasmin has recently been shown to function as an antioxidant in neutralizing nitric oxide via nitrosothiol formation (Inoue et al. 1999). A membrane-bound (GPI anchor) form of ceruloplasmin has been found in glia cells, which might have cytoprotective function in brain (Patel and David 1997). Lack of that form of ceruloplasmin in the aceruloplasminemic patient could also result in neurodegeneration.

Molecular genetic studies of DNA from cells from aceruloplasminemic individuals have detected specific mutations in the ceruloplasmin mRNA. Mutations that occur often lead to splicing errors and abridged forms of ceruloplasmin that cannot support normal ferroxidase function (Yazaki et al. 1998). In one patient with the disease, a 5-bp insertion in exon 7 caused an out-of-frame shift that generated a stop codon that aborted the protein from the ribosome before synthesis was completed (Harris et al. 1995). In another, a guanine-to-adenine transition at a splice acceptor site caused a similar premature termination of the protein during biosynthesis (Yoshida et al. 1995).

Despite careful clinical investigations of afflicted individuals, it is still unclear how ceruloplasmin controls iron homeostasis and tissue distribution. Copper-deficiency studies in which ceruloplasmin is lowered never achieve a total absence of the protein, as is observed in aceruloplasminemia. Such studies are further confounded by a disrupted copper status. A mouse knockout model, however, has recently been developed, and preliminary studies show no abnormalities in cellular iron uptake but do show a pronounced impairment in the movement of iron out of the reticuloendothelial cells and hepatocytes (Harris et al. 1998). An issue that merits investigation is the extent to which aceruloplasminemia might modulate an individual's susceptibility to copper toxicity.

TYROLEAN INFANTILE CIRRHOSIS

Between about 1900 and 1980, 138 infants and young children died in the Tyrolean area of western Austria from liver cirrhosis; that syndrome has been termed Tyrolean infantile cirrhosis (TIC) (Müller et al. 1996). The pathology of TIC is indistinguishable from Indian childhood cirrhosis and other forms of hepatic copper toxicosis (see descriptions below) (Müller et al. 1996). A common feature among the infants and young children with TIC was having been fed a one-to-one mixture of unpasteurized cow's milk with water heated for about 20 min in old copper pots (Müller et al. 1996). Such a preparation results in a high concentration of copper in the milk (Müller et al. 1996). Copper cooking utensils were common in the Tyrolean area because of an extensive copper mining industry that was present until about 1926. The disease appears to be an autosomal recessive disorder (Müller et al. 1996), whose manifestations might be due to a combination of a genetic predisposition and a high intake of copper. Reports of this syndrome have essentially disappeared since about 1980, probably as copper cooking utensils were gradually replaced in the region by modern cooking utensils in the mid-to-late 1960s.

INDIAN CHILDHOOD CIRRHOSIS

The etiology of Indian childhood cirrhosis (ICC) appears to be similar to that of TIC (Bavdekar et al. 1996; Bhave et al. 1982; Pandit and Bhave 1996; Popper et al. 1979; Tanner et al. 1983; Tanner et al. 1979). ICC occurs in India in infants and very young children fed milk stored in brass or copper containers. Copper might have a role in the disease for the following reasons: milk stored in that manner is very high in copper (Popper et al. 1979; Tanner et al. 1979; Bhave et al. 1982; Tanner et al. 1983; Bavdekar et al. 1996; Pandit and Bhave 1996); high copper concentrations have been found in the livers of ICC patients (Bavdekar et al. 1996; Bhave et al. 1982; Pandit and Bhave 1996; Popper et al. 1979; Tanner et al. 1983; Tanner et al. 1979); penicillamine treatment rapidly improves the symptoms (Bavdekar et al. 1996; Bhave et al. 1982; Pandit and Bhave 1996; Popper et al. 1979; Tanner et al. 1983; Tanner et al. 1979); and other possible causes of liver disease have been ruled out in most of the patients (Bavdekar et al. 1996; Bhave et al. 1982; Pandit and Bhave 1996; Popper et al. 1979; Tanner et al. 1983; Tanner et al. 1979). In addition, once storage of milk in copper and brass containers was reduced, the disease began to disappear. An autosomal recessive component also seems to play a role in the disease, because the siblings, but not the parents, of ICC patients were affected (Bavdekar et al. 1996; Bhave et al. 1982; Pandit and Bhave 1996; Popper et al. 1979; Tanner et al. 1983; Tanner et al. 1979). Therefore, as with TIC, the liver disease in ICC might be due to genetic abnormalities in copper metabolism and a high copper intake.

IDIOPATHIC COPPER TOXICOSIS

Cases of idiopathic copper toxicosis (ICT) have been reported in numerous countries. Those cases may arise from a number of different etiologies. ICT has also been called ICC-like cirrhosis, copper-associated childhood cirrhosis, and copper-associated liver disease in childhood (Lefkowitch et al. 1982; Adamson et al. 1992; Aljajeh et al. 1994; Horslen et al. 1994; Müller et al. 1998). ICT has been reviewed recently by Müller et al. (1998). Most ICT cases are infants or very young children, but some are children up to 10 years of age. In many cases, increased ingestion of copper has been demonstrated, and there is either consanguinity of parents or involvement of more than one sibling. Müller et al. (1998) hypothesized that ICT is caused by a combination of an autosomal recessive inherited defect in copper metabolism and excess copper intake. In some cases, especially in infants and toddlers, the source of excess copper appears to have been drinking water (Müller et al. 1998) (see Chapter 5).

OTHER GENETIC DISORDERS

Two additional genetic conditions of humans that might enhance the susceptibility to copper toxicosis deserve brief mention. Individuals with a rare type of alloalbuminemia, "albumin Christchurch," have a mutation in the gene that encodes the primary copper-binding site of plasma albumin, and the albumin Christchurch has reduced ability to bind Cu(II) compared with normal albumin. (Brennan and Carrell 1980). As a consequence, individuals with albumin Christchurch have functional impairment in plasma capacity to transport ionic copper. Individuals with type I-A oculocutaneous albinism have mutations in the gene that encodes tyrosinase, a key copper-containing enzyme in the pathway of melanin biosynthesis (Oetting and King 1992; Passmore et al. 1999). The melanin of pigmented tissues (e.g., retina and skin) avidly binds copper, zinc, and other metal ions in vitro (Froncisz et al 1980; Andrzejczyk and Buszman 1992) and in vivo (Bowness and Morton 1953; Horcicko et al. 1973). Melanin is postulated as a possible storage repository for essential metals (Pfeiffer and Mailloux 1988), and indications of abnormal metabolism of copper and zinc have been reported in black albino patients, compared with Caucasian albino patients and control groups (Silverstone et al 1986). Subjects with oculocutaneous albinism lack tyrosinase activity. That deficiency is associated with an absence of melanin, which can sequester and detoxify copper and other metals. Therefore, patients with oculocutaneous albinism might be prone to developing copper toxicosis.

DISEASE-INDUCED CHANGES IN COPPER HOMEOSTASIS

In addition to the above genetic disorders, there are a number of environmental and physiological conditions that can perturb copper metabolism and alter the delivery of copper to select tissues. The conditions shown to alter copper metabolism are diverse and include exercise, infection, inflammation, cirrhosis, diabetes, and hypertension (Sandstead 1995; Disilvestro et al. 1992; Turnlund 1988; Walter et al. 1991; Miesel and Zuber 1993; Keen 1993). In all of those conditions, a common finding is hypercupremia, which is due to high concentrations of plasma ceruloplasmin. The increase in plasma ceruloplasmin is the result of an increase in the synthesis of ceruloplasmin in the liver, with its subsequent release into the plasma pool. An increase in ceruloplasmin synthesis represents one component of the so-called acute-phase response. This response is triggered by cytokines and select hormones that are released in response to tissue injury (Cousins 1985; Dinarello 1989). Although the acute-phase response is transitory in most cases, it can persist for long periods if tissue injury is continuous. The long-term health consequences of hypercupremia have not been well defined; however, some investigators have reported that the occurrence of hypercupremia is a risk factor for cardiovascular disease and for some cancers (Salonen et al. 1991). Mechanistically, it can be argued that persistent hypercupremia represents an oxidative challenge, if the hypercupremia is associated with a rise in low-molecular-bound copper pools, which have the potential to be involved in redox cycling and, thus, the generation of reactive oxygen species. An increase in DNA damage, secondary to copper-induced oxidative stress, could present a carcinogenic risk. Similarly, an increase in oxidative stress could result in an increased risk for some forms of cardiovascular disease (Dubick et al. 1999; Esterbauer et al. 1992; Gey et al. 1991). Given the above, it can be hypothesized that chronic hypercupremia can be a causative factor in the initiation of some diseases. An alternative argument to the above is that the hypercupremia is simply a marker for either early tissue damage or other metabolic abnormalities that are the primary initiators of the vascular disease or cancer. Most investigators think that hypercupremia is an effect rather than a cause of disease. Additional research on this issue is needed.

It is important to point out that the occurrence of high concentrations of copper in plasma does not necessarily translate into high concentrations in soft tissue; indeed, high plasma concentrations in some diseases, including diabetes and hypertension, have been correlated with low concentrations in muscle, liver, and aortic tissue (Dubick et al. 1999; Sjogren et al. 1986; Tilson 1982). With respect to hypertension, the low copper concentrations in aortic tissue have been correlated with low activity in copper zinc superoxide dismutase and high concentrations of tissue lipid peroxide (Dubick et al. 1987; Dubick et al. 1999). The above observations are consistent with the concept that copper deficiency is a risk factor for cardiovascular disease (Klevay 1989), although the copper deficiency occurs due to disease-induced changes in copper metabolism rather than to dietary copper deficiency.

It is not known to what extent disease-induced changes in copper metabolism alter the requirement for dietary copper or the sensitivity to copper toxicity. However, if persistent hypercupremia results in an increased excretion of copper, this condition might result in an increased dietary requirement for copper.

CONCLUSIONS

- Menkes disease is X linked, and Wilson disease is an autosomal recessive disorder.
- With the recent identification of the genes responsible for Menkes and Wilson diseases, considerable progress has been made in the understanding of human copper metabolism.
- With the identification of the Menkes and Wilson genes, rapid progress has been made in the understanding of the pathophysiology, pathogenesis, and diagnosis of these disorders.
- Heterozygous carriers of the Wilson-disease gene can have abnormal copper metabolism and, consequently, excess accumulation of copper.
- TIC, ICC, and ICT occur in infants or young children. TIC and ICC appear to have familial components with an autosomal recessive inheritance pattern. Some cases of ICT appear to have an autosomal recessive

inheritance pattern, and others do not. In all three disorders, excess copper ingestion is typically found.
- A number of diseases can result in chronic hypercupremia.

RECOMMENDATIONS

- Research should be conducted to establish the frequency and characteristics of the Wilson-disease gene defects.
- The influence of heterozygosity for gene defects that alter copper transport or susceptibility to copper toxicity needs to be defined.
- Research should be conducted to establish the genetic causes of susceptibility to copper in TIC, ICC and ICT and the frequency of these genetic defects.
- The potential risk of copper toxicity in patients with aceruloplasminemia needs to be determined.

REFERENCES

Adamson, M., B. Reiner, J.L. Olson, Z. Goodnam, L. Plotnick, I. Bernardini, and W.A. Gahl. 1992. Indian childhood cirrhosis in an American child. Gastroenterology 102(5):1771-1777.

Aljajeh, A., S. Mughal, B. al-Tahou, T. Ajrawi, E.A. Ismail, and N.C. Nayak. 1994. Indian childhood cirrhosis-like liver disease in an Arab child. A brief report. Virchows Arch. 424(2):225-227.

Andrzejczyk, J., and E. Buszman. 1992. Interaction of Fe^{3+}, Cu^{2+} and Zn^{2+} with melanin and melanoproteins from bovine eyes. Acta Biochim. Pol. 39(1):85-88.

Bachmann, H., J. Lössner, B. Gruss, and U. Ruchholtz. 1979. The epidemiology of Wilson's disease in the German Democratic Republic and current problems from the viewpoint of population genetics [in German]. Psychiatr. Neurol. Med. Psychol. 31(7):393-400.

Bavdekar, A.R., S.A. Bhave, A.M. Pradhan, A.N. Pandit, and M.S. Tanner. 1996. Long term survival in Indian childhood cirrhosis treated with D-penicillamine. Arch. Dis. Child. 74(1):32-35.

Bhave, S.A., A.N. Pandit, A.M. Pradhan, D.G. Sidhaye, A. Kantarjian, A. Williams, I.C. Talbot, and M.S. Tanner. 1982. Liver disease in India. Arch. Dis. Child. 57(12):922-928.

Bowness, J.M., and R.A. Morton. 1953. The association of zinc and other metals with melanin and melanin-protein complex. Biochem. J. 53(4): 620-626.

Brennan, S.O. and R.W. Carrell. 1980. Functional abnormality of proalbumin Christchurch. Biochim. Biophys. Acta 621(1):83-88.

Brewer, G.J. In press. Editorial: Is heterozygosity for a Wilson's disease gene defect an important underlying cause of infantile and childhood copper toxicosis syndromes? J. Trace Elem. Exp. Med.

Brewer, G.J. and V. Yuzbasiyan-Gurkan. 1992. Wilson disease. Medicine 71(3):139-164.

Brewer, G.J., R.D. Dick, V. Johnson, J.A. Brunberg, K.J. Kluin, and J.K. Fink. 1998. Treatment of Wilson's disease with zinc: XV. Long-term follow-up studies. J. Lab. Clin. Med. 132(4):264-278.

Bull, P.C., G.R. Thomas, J.M. Rommens, J.R. Forbes, and D.W. Cox. 1993. The Wilson disease gene is a putative copper transporting P-type ATPase similar to the Menkes' gene. Nat. Genet. 5(4):327-337.

Bush, J.A., J.P. Mahoney, H. Markowitz, C.J. Gubler, G.E. Cartwright, and M.M. Wintrobe. 1955. Studies on copper metabolism. XVI. Radioactive copper studies in normal subjects and in patients with hepatolenticular degeneration. J. Clin. Invest. 34:1766-1778.

Camakaris, J., M.J. Petris, L. Bailey, P. Shen, P. Lockhart, T.W. Glover, C. Barcroft, J. Patton, and J.F. Mercer. 1995. Gene amplification of the

Menkes (MNK; ATP7A) P-type ATPase gene of CHO cells is associated with copper resistance and enhanced copper efflux. Hum. Mol. Genet. 4(11):2117-2123.

Cecchi, C., and P. Avner. 1996. Genomic organization of the mottled gene, the mouse homologue of the human Menkes disease gene. Genomics 37(1):96-104.

Chelly, J., Z. Tumer, T. Tonnesen, A. Petterson, Y. Ishikawa-Brush, N. Tommerup, N. Horn, and A.P. Monaco. 1993. Isolation of a candidate gene for Menkes disease that encodes a potential heavy metal binding protein. Nat. Genet. 3(1):14-19.

Chowrimootoo, G.F., H.A. Ahmed, and C.A. Seymour. 1996. New insights into the pathogenesis of copper toxicity in Wilson's disease: Evidence for copper incorporation and defective canalicular transport of caeruloplasmin. Biochem. J. 315(Pt 3):851-855.

Christodoulou, J., D.M. Danks, B. Sarkar, K.E. Baerlocher, R. Casey, N. Horn, Z. Tumer, and J.T. Clarke. 1998. Early treatment of Menkes disease with parenteral copper-histidine: Long-term follow-up of four treated patients. Am. J. Med. Genet. 76(2):154-164.

Cousins, R.J. 1985. Absorption, transport and hepatic metabolism of copper and zinc. Physiol. Rev. 65(2):238-309.

Cox, D.W., and E.A. Roberts. 1999. Wilson Disease. GeneClinics, University of Washington, Seattle. Online. Available: wysiwyg://79/http://www.geneclinics.org/profiles/wilson/details.html

Danks, D.M. 1995. Disorders of copper transport. Pp. 2211-2235 in The Metabolic and Molecular Basis of Inherited Disease, 7th Ed., C.R. Scriver, A.L. Beaudet, W.M. Sly, and D. Valle, eds. New York: McGraw-Hill.

Danks, D.M., P.E. Campbell, B.J. Stevens, V. Mayne, and E. Cartwright. 1972. Menkes's kinky hair syndrome. An inherited defect in copper absorption with widespread effects. Pediatrics 50(2):188-201.

Das, S., B. Levinson, C. Vulpe, S. Whitney, J. Gitschier, and S. Packman S. 1995. Similar splicing mutations of the Menkes/mottled copper-transporting ATPase gene in occipital horn syndrome and the blotchy mouse. Am. J. Hum. Genet. 56(3):570-576.

Davis, W., G.F. Chowrimootoo, and C.A. Seymour. 1996. Defective biliary excretion in Wilson's disease: The role of caeruloplasmin. Eur. J. Clin. Invest. 26(10):893-901.

De Paepe, A., B. Loeys, K. Devriendt, and J.P. Fryns. 1999. Occipital Horn syndrome in a 2-year-old boy. Clin. Dysmorphol. 8(3):179-183.

Dierick, H.A., A.N. Adam, J.F. Escara-Wilke, and T.W. Glover. 1997. Immunocytochemical localization of the Menkes copper transport protein (ATP7A) to the trans-Golgi network. Hum. Mol. Genet. 6(3): 409-416.

Dierick, H.A., L. Ambrosini, J. Spencer, T.W. Glover, and J.F. Mercer. 1995. Molecular structure of the Menkes disease gene (ATP7A). Genomics 28(3):462-469.
Dinarello, C.A. 1989. The endogenous pyrogens in host-defense interactions. Hosp. Pract. (Off Ed) 24(11):111-5, 118, 121 passim.
DiSilvestro, R.A., J. Marten, and M. Skehan. 1992. Effects of copper supplementation on ceruloplasmin and copper-zinc superoxide dismutase in free-living rheumatoid arthritis patients. J. Am. Coll. Nutr. 11(2):177-80.
Dubick, M.A., G.C. Hunger, S.M. Casey, and C.L. Keen. 1987. Aortic ascorbic acid, trace elements, and superoxide dismutase activity in human aneurysmal and occlusive disease. Proc. Soc. Exp. Biol. Med. 184(2):138-143.
Dubick, M.A., C.L. Keen, R.A. DiSilvestro, C.D. Eskelson, J. Ireton, and G.C. Hunter. 1999. Antioxidant enzyme activity in human abdominal aortic aneurysmal and occlusive disease. Proc. Soc. Exp. Biol. Med. 220(1):39-45.
Esterbauer, H., J. Gebicki, H. Puhl, and G. Jurgens. 1992. The role of lipid peroxidation and antioxidants in oxidative modification of LDL. Free Radic. Biol. Med. 13(4):341-390.
Francis, M.J., E.E. Jones, E.R. Levy, S. Ponnambalam, J. Chelly, and A.P. Monaco. 1998. A Golgi localization signal identified in the Menkes recombinant protein. Hum. Mol. Genet. 7(8):1245-1252.
Frommer, D.J. 1974. Defective biliary excretion of copper in Wilson's disease. Gut 15(2):125-129.
Froncisz, W., T. Sarna, and J.S. Hyde. 1980. Cu^{2+} probe of metal-ion binding sites in melanin using electron paramagnetic resonance spectroscopy. I. Synthetic melanins. Arch. Biochem. Biophys. 202(1):289-303.
Gey, K.F., P. Puska, P. Jordan, and U.K. Moser. 1991. Inverse correlation between plasma vitamin E and mortality from ischemic heart disease in cross-cultural epidemiology. Am. J. Clin. Nutr. 53(1 Suppl):326S-334S.
Giagheddu, A, L. Demelia, G. Puggioni, A.M. Nurchi, L. Contu, G. Pirari, A. Deplano, and M.G. Rachele. 1985. Epidemiologic study of hepatolenticular degeneration (Wilson's disease) in Sardinia (1902-1983). Acta Neurol. Scand. 72(1):43-45.
Gibbs, K., and J.M. Walshe. 1980. Biliary excretion of copper in Wilson's disease. Lancet 2(8193):538-539.
Gitlin, J.D. 1998. Aceruloplasminemia. Pediatr. Res. 44(3):271-276.
Goka, T.J., R.E. Stevenson, P.M. Hefferan, and R.R. Howell. 1976. Menkes disease: A biochemical abnormality in cultured human fibroblasts. Proc. Natl. Acad. Sci. (USA) 73(2):604-606.
Harris, Z.L., L.W. Klomp and J.D. Gitlin. 1998. Aceruloplasminemia: an

inherited neurodegenerative disease with impairment of iron homeostasis. Am. J. Clin. Nutr. 67(5 Suppl):972S-977S.

Harris, Z.L., Y. Takahashi, H. Miyajima, M. Serizawa, R.T. MacGillivray, and J.D. Gitlin. 1995. Aceruloplasminemia: Molecular characterization of this disorder of iron metabolism. Proc. Natl. Acad. Sci. (USA) 92(7):2539-2543.

Hill, G.M., G.J. Brewer, A.S. Prasad, C.R. Hydrick, and D.E. Hartmann. 1987. Treatment of Wilson's disease with zinc: I. Oral zinc therapy regimens. Hepatology 7(3):522-528.

Holden, J.M., W.R. Wolf, and W. Mertz. 1979. Zinc and copper in self selected diets. J. Am. Diet Assoc. 75(1):23-28.

Horcicko, J., J. Borovansky, J. Duchon and B. Prochazkova. 1973. Distribution of zinc and copper in pigmented tissues. Hoppe Seylers Z. Physiol. Chem. 354(2):203-204.

Horn, N. 1976. Copper incorporation studies on cultured cells for prenatal diagnosis of Menkes' disease. Lancet 1(7970):1156-8.

Horn, N., T. Tonnesen, and Z. Tumer. 1992. Menkes disease: An X-linked neurological disorder of the copper metabolism. Brain Pathol. 2(4):351-362.

Horslen, S.P., M.S. Tanner, T.D. Lyon, G.S. Fell, and M.F. Lowry. 1994. Copper associated childhood cirrhosis. Gut 35(10):1497-1500.

Inoue, K., T. Akaike, Y. Miyamoto, T. Okamoto, T. Sawa, M. Otagiri, S. Suzuki, T. Yoshimura, and H. Maeda. 1999. Nitrosothiol formation catalyzed by ceruloplasmin. Implication for cytoprotective mechanism in vivo. J. Biol. Chem. 274(38):27069-27075.

Iyengar, V, G.J. Brewer, R.D. Dick, and O.Y. Chung. 1988. Studies of cholecystokinin-stimulated biliary secretions reveal a high molecular weight copper-binding substance in normal subjects that is absent in patients with Wilson's disease. J. Lab. Clin. Med. 111(3):267-274.

Kaler, S.G., S. Das, B. Levinson, D.S. Goldstein, C.S. Holmes, N.J. Patronas, S. Packman, and W.A. Gahl. 1996. Successful early copper therapy in menkes disease associated with a mutant transcript containing a small In-frame deletion. Biochem. Mol. Med. 57(1):37-46.

Keen, C.L. 1993. Efects of exercise and heat on mineral metabolism and requirements. Pp. 117-135 in Nutritional Needs in Hot Environments. B.M. Marriott, ed. Washington, DC: National Academy Press.

Klevay, L.M. 1989. Ischemic heart disease as copper deficiency. Adv. Exp. Med. Biol. 258:197-208.

Klevay, L.M., S.J. Reck, and D.F. Barcome. 1979. Evidence of dietary copper and zinc deficiencies. JAMA 241(18):1916-1918.

Kodama, H. 1993. Recent developments in Menkes disease. J. Inherit. Metab. Dis. 16(4):791-799.

Kodama, H., and Y. Murata. 1999. Molecular genetics and pathophysiology of Menkes disease. Pediatr. Int. 41(4):430-435.

Kodama, H., Y. Murata, and M. Kobayashi. 1999. Clinical manifestations and treatment of Menkes disease and its variants. Pediatr. Int. 41(4): 423-429.

Lefkowitch, J.H., C.L. Honig, M.E. King, and J.W. Hagstom. 1982. Hepatic copper overload and features of Indian childhood cirrhosis in an American sibship. N. Engl. J. Med. 307(5):271-277.

Levinson, B., R. Conant, R. Schnur, S. Das, S. Packman, J. Gitschier. 1996. A repeated element in the regulatory region of the MNK gene and its deletion in a patient with occipital horn syndrome. Hum. Mol. Genet. 5(11):1737-42.

Logan, J.I., K.B. Harveyson, G.B. Wisdom, A.E. Hughes and G.P. Archbold. 1994. Hereditary caeruloplasmin deficiency, dementia and diabetes mellitus. QJM 87(11):663-70.

Loudianos, G., V. Dessi, M. Lovicu, A. Angius, A. Figus, F. Lilliu, S. De Virgiliis, A.M. Nurchi, A. Deplano, P. Moi, M. Pirastu, and A. Cao. 1999. Molecular characterization of wilson disease in the Sardinian population—Evidence of a founder effect. Hum. Mutat. 14(4):294-303.

Lutsenko, S., and M.J. Cooper. 1998. Localization of the Wilson's disease protein product to mitochondria. Proc. Natl. Acad. Sci. (USA) 95(11): 6004-6009.

Masson, W., H. Hughes, D. Papworth, Y. Boyd and N. Horn. 1997. Abnormalities of copper accumulation in cell lines established from nine different alleles of mottled are the same as those found in Menkes disease. J. Med. Genet. 34(9):729-732.

Menkes, J.H., M. Alter, G.K. Steigleder, D.R. Weakley, and J.H. Sung. 1962. A sex-linked recessive disorder with retardation of growth, peculiar hair, and focal cerebral and cerebellar degeneration. Pediatrics 29:764-779.

Mercer, J.F., J. Livingston, B. Hall, J.A. Paynter, C. Begy, S. Chandrasekharappa, P. Lockhart, A. Grimes, M. Bhave, and D. Siemieniak. 1993. Isolation of a partial candidate gene for Menkes disease by positional cloning. Nat. Genet. 3(1):20-25.

Miesel, R, and M. Zuber. 1993. Copper-dependent antioxidase defenses in inflammatory and autoimmune rheumatic diseases. Inflammation 17(3):283-294.

Milne, D.B. 1998. Copper intake and assessment of copper status. Am. J. Clin. Nutr. 67(5 Suppl.):1041S-1045S.

Miyajima, H., M. Fujimoto, S. Kohno, E. Kaneko, and J.D. Gitlin. 1998. CSF abnormalities in patients with aceruloplasminemia. Neurology 51(4):1188-1190.

Miyajima, H., S. Kohno, Y. Takahashi, O. Yonekawa and T. Kanno. 1999. Estimation of the gene frequency of aceruloplasminemia in Japan. Neurology 53(3):617-619.

Miyajima, H, Y. Nishimura, K. Mizoguchi, M. Sakamoto, T. Shimizu, and

N. Honda. 1987. Familial apoceruloplasmin deficiency associated with blepharospasm and retinal degeneration. Neurology 37(5):761-767.

Miyajima, H., Y. Takahashi, M. Serizawa, E. Kaneko, and J.D. Gitlin. 1996. Increased plasma lipid peroxidation in patients with aceruloplasminemia. Free Radic. Biol. Med. 20(5):757-760.

Müller, T, H. Feichtinger, H. Berger, and W. Müller. 1996. Endemic Tyrolean infantile cirrhosis: An exogenetic disorder. Lancet 347(9005):877-880.

Müller, T, W. Müller, and H. Feichtinger. 1998. Idiopathic copper toxicosis. Am. J. Clin. Nutr. 67(5 Suppl.):1082S-1086S.

Oetting, W.S., and R.A. King. 1992. Molecular analysis of type I-A (tyrosinase negative) oculocutaneous albinism. Hum. Genet. 90(3):258-262.

Oh, W.J., E.K. Kim, K.D. Park, S.H. Hahn, and O.J. Yoo. 1999. Cloning and characterization of the promoter region of the Wilson disease gene. Biochem. Biophys. Res. Commun. 259(1):206-211.

O'Reilly, S., P.M. Weber, M. Oswald and L. Shipley. 1971. Abnormalities of the physiology of copper in Wilson's disease. 3. The excretion of copper. Arch. Neurol. 25(1):28-32.

Pandit, A., and S. Bhave. 1996. Present interpretation of the role of copper in Indian childhood cirrhosis. Am. J. Clin. Nutr. 63:930S-835S.

Passmore, L.A., B. Kaesmann-Kellner, B.H.F. Weber. 1999. Novel and recurrent mutations in the tyrosinase gene and the P gene in the German albino population. Hum. Genet. 105(3):200-210.

Patel, B.N., and S. David. 1997. A novel glycosylphosphatidylinositol-anchored form of ceruloplasmin is expressed by mammalian astrocytes. J. Biol. Chem. 272(32):20185-190.

Petrukhin, K., S.G. Fischer, M. Pirastu, R.E. Tanzi, I. Chernov, M. Devoto, L.M. Brzustowicz, E. Cayanis, E. Vitale, and J.J. Russo. 1993. Mapping, cloning and genetic characterization of the region containing the Wilson disease gene. Nat. Genet. 5(4):338-343.

Pfeiffer, C.C. R.J. Mailloux. 1988. Hypertension: heavy metals, useful cations and melanin as a possible repository. Med. Hypotheses 26(2):125-130.

Popper, H., S. Goldfischer, I. Sternlieb, N.C. Nayak and T.V. Madhavan. 1979. Cytoplasmic copper and its toxic effects. Studies in Indian childhood cirrhosis. Lancet 1(8128):1205-1208.

Proud, V.K., H.G. Mussell, S.G. Kaler, D.W. Young, and A.K. Percy. 1996. Distinctive Menkes disease variant with occipital horns: delineation of natural history and clinical phenotype. Am. J. Med. Genet. 65(1):44-51.

Qi, M., and P.H. Byers. 1998. Constitutive skipping of alternatively spliced exon 10 in the ATP7A gene abolishes Golgi localization of the Menkes protein and produces the occipital horn syndrome. Hum. Mol. Genet. 7(3):465-469.

Qian, Y., E. Tiffany-Castiglioni, and E.D. Harris. 1997. A Menkes P-type ATPase involved in copper homeostasis in the central nervous system of the rat. Brain Res. Mol. Brain. Res. 48(1):60-66.
Qian, Y., E. Tiffany-Castiglioni, J. Welsh, and E.D. Harris. 1998. Copper efflux from murine microvascular cells requires expression of the Menkes disease Cu-ATPase. J. Nutr. 128(8):1276-1282.
Rayner, M.H. and K.T. Suzuki. 1994. Effect of medium copper concentration on the growth, uptake and intracellular balance of copper and zinc in Menkes' and normal control cells. Biometals 7(3):253-260.
Reilly, M., L. Daly and M. Hutchinson. 1993. An epidemiological study of Wilson's disease in the Republic of Ireland. J. Neurol. Neurosurg. Psychiatry 56(3):298-300.
Reiser, S., J.C. Smith Jr, W. Mertz, J.T. Holbrook, D.J. Scholfield, A.S. Powell, W.K. Canfield and J.J. Canary. 1985. Indices of copper status in humans consuming a typical American diet containing either fructose or starch. Am. J. Clin. Nutr. 42(2):242-251.
Royce, P.M., J. Camakaris, J.R. Mann, and D.M. Danks. 1982. Copper metabolism in mottled mouse mutants. The effect of copper therapy on lysyl oxidase activity in brindled (Mobr) mice. Biochem. J. 202(2):369-371.
Salonen, J.T., R. Salonen, H. Korpela, S. Suntioinen, and J. Tuomilehto. 1991. Serum copper and the risk of acute myocardial infarction: A prospective population study in men in eastern Finland. Am. J. Epidemiol. 134(3):268-276.
Sandstead, H.H. 1995. Requirements and toxicity of essential trace elements, illustrated by zinc and copper. Am. J. Clin. Nutr. 61(3 Suppl): 621S-624S.
Sarkar, B., K. Lingertat-Walsh and J.T. Clarke. 1993. Copper-histidine therapy for Menkes disease. J. Pediatr. 123(5):828-830.
Scheinberg, I., H. and I. Sternlieb. 1984. Wilson's disease. Major Problems in Internal Medicine, Vol. 23, L.H. Smith, ed. Philadelphia: W.B. Saunders.
Silver, S., G. Nucifora, and L.T. Phung. 1993. Human Menkes X-chromosome disease and the staphylococcal cadmium-resistance ATPase: A remarkable similarity in protein sequences. Mol. Microbiol. 10(1):7-12.
Silverstone, B.Z., I. Nawratzki, D. Berson and L. Yanko. 1986. Zinc and copper metabolism in oculocutaneous albinism in the Caucasian. Metab. Pediatr. Syst. Ophthalmol. 9(1):589-591.
Sjögren, A., L. Edvinsson, C.H. Florén, and M. Abdulla. 1986. Zinc and copper in striated muscle and body fluids from subjects with diabetes mellitus type I. Nutr. Res. 6(2):147-154.
Solioz, M., and C. Vulpe. 1996. Cpx-type ATPases: A class of P-type ATPases that pump heavy metals. Trends. Biochem. Sci. 21(7):237-241.
Strickland, G.T., W.M. Beckner, M.L. Leu, and S. O'Reilly. 1969. Copper-

67 studies in Wilson's disease patients and their families. Clin. Res. 17(2):396.
Tanaka, K., K. Kobayashi, Y. Fujita, C. Fukuhara, S. Onosaka, and K. Min. 1990. Effects of chelators on copper therapy of macular mouse, a model animal of Menkes' kinky disease. Res. Commun. Chem. Pathol. Pharmacol. 69(2):217-227.
Tanner, M.S., A.H. Kantarjian, S.A. Bhave, and A.N. Pandit. 1983. Early introduction of copper-contaminated animal milk feeds as a possible cause of Indian childhood cirrhosis. Lancet 2(8357):992-995.
Tanner, M.S., B. Portmann, A.P. Mowat, R. Williams, A.N. Pandit, C.F. Mills, and I. Bremner. 1979. Increased hepatic copper concentration in Indian childhood cirrhosis. Lancet 1(8128):1203-1205.
Tanzi, R.E., K. Petrukhin, J.L. Pellequer, W. Wasco, B. Ross, D.M. Romano, E. Parano, L. Pavone, and L.M. Brzustowicz. 1993. The Wilson disease gene is a copper transporting ATPase with homology to the Menkes disease gene. Nat. Genet. 5(4):44-50.
Terada, K., and T. Sugiyama. 1999. The Long-Evans Cinnamon rat: an animal model for Wilson's disease. Pediatr. Int. 41(4):414-418.
Thomas, G.R., J.R. Forbes, E.A. Roberts, J.M. Walshe, and D.W. Cox. 1995. The Wilson disease gene: Spectrum of mutations and their consequences. Nat. Genet. 9(2):210-7.
Tilson, M.D. 1982. Decreased hepatic copper levels. A possible chemical marker for the pathogenesis of aortic aneurysms in man. Arch. Surg. 117(9):1212-1213.
Tumer, Z., B. Vural, T. Tonnesen, J. Chelly, A.P. Monaco, and N. Horn. 1995. Characterization of the exon structure of the Menkes disease gene using vectorette PCR. Genomics 26(3):437-442.
Turnlund, J.R. 1988. Copper nutriture, bioavailability, and the influence of dietary factors. J. Am. Diet Assoc. 88(3):303-308.
Voskoboinik I, H. Brooks, S. Smith, P. Shen, and J. Camakaris. 1998. ATP-dependent copper transport by the Menkes protein in membrane vesicles isolated from cultured Chinese hamster ovary cells. FEBS Lett 435(2-3):178-182.
Voskoboinik, I., D. Strausak, M. Greenough, H. Brooks, M. Petris, S. Smith, J.F. Mercer, and J. Camakaris. 1999. Functional analysis of the N-terminial CXXC metal-binding motifs in the human Menkes copper-transporting P-type ATPase expressed in cultured mammalian cells. J. Biol. Chem. 274(31):22008-22012.
Vulpe, C.D., and S. Packman. 1995. Cellular copper transport. Annu. Rev. Nutr. 15:293-322.
Vulpe, C, B. Levinson, S. Whitney, S. Packman and J. Gitschier. 1993. Isolation of a candidate gene for Menkes disease and evidence that it encodes a copper-transporting ATPase. Nat. Genet. 3(1):7-13.
Walter, R.M. Jr, J.Y. Uriu-Hare, K.L. Olin, M.H. Oster, B.D. Anawalt,

J.W. Critchfield, and C.L. Keen. 1991. Copper, zinc, manganese, and magnesium status and complications of diabetes mellitus. Diabetes Care 14(11):1050-1056.

Yamaguchi, Y, M.E. Heiny and J.D. Gitlin. 1993. Isolation and characterization of a human liver cDNA as a candidate gene for Wilson's disease. Biochem. Biophys. Res. Commun. 197(1):271-277.

Yazaki, M., K. Yoshida, A. Nakamura, K. Furihata, M. Yonekawa, T. Okabe, N. Yamashita, M. Ohta and S. Ikeda. 1998. A novel splicing mutation in the ceruloplasmin gene responsible for hereditary ceruloplasmin deficiency with hemosiderosis. J. Neurol. Sci. 156(1):30-34.

Yoshida, K., K. Furihata, S. Takeda, A. Nakamura, K. Yamamoto, H. Morita, S. Hiyamuta, S. Ikeda, N. Shimizu, and N. Yanagisawa. 1995. A mutation in the ceruloplasmin gene is associated with systemic hemosiderosis in humans. Nat. Genet. 9(3):267-72.

Yoshimura, N., K. Kida, S. Usutani, and M. Nishimura. 1995. Histochemical localization of copper in various organs of brindled mice after copper therapy. Pathol. Int. 45(1):10-18.

Yuzbasiyan-Gurkan, V, V. Johnson, and G.J. Brewer. 1991. Diagnosis and characterization of presymptomatic patients with Wilson's disease and the use of molecular genetics to aid in the diagnosis. J. Lab. Clin. Med. 118(5):458-465.

5

Health Effects of Excess Copper

THIS chapter focuses on the health effects associated with acute and chronic exposure to excess copper. Information on those effects comes from human case-reports and population-based studies. The emphasis is placed on acute exposure effects on the gastrointestinal (GI) system. Effects on other target organs, such as the liver, in subjects following high-dose chronic exposure and in sensitive populations are considered. Toxicity data from animal studies are presented, and the use of animal models for studying the mechanism underlying the toxicity of copper in humans is discussed.

ACUTE TOXICITY

Case Reports and Population-Based Studies

Human cases of acute copper toxicosis are presented in this section. The cases are cited in reports by the NRC (1977), EPA (1987), the U.S. Agency for Toxic Substances and Disease Registry (ATSDR 1990), and the World Health Organization's International Programme of Chemical Safety (IPCS 1998). Table 5-1 summarizes the reported health effects of ingested copper in humans. Most human data on high-dose acute poisoning are based on cases of suicidal intent with the ingestion of copper compounds or accidental consumption of copper-contaminated foods and beverages. In such cases, it is difficult to estimate the quantity of copper consumed, whether it was in solid form, aqueous suspension, or solution. It is also

TABLE 5-1 Case Reports of Copper Toxicosis Following Oral Exposures of Humans to Copper Salts

Reference	Sex, Age, Number	Exposures	Clinical Sequelae and Comments
Percival 1784	F, 17 yr, 1	Ingestion of ≈100 to 130 g of a pickled vegetable, shown by analysis to be contaminated with Cu	Abdominal pain, generalized rash, flatulence, vomiting, diarrhea, cardiac arrhythmia, death on the 8th day post-exposure
Griffin 1951	M, adults, 6	Ingestion of coffee brewed with water from a Cu-contaminated, gas-heated boiler	Nausea, vomiting, malaise
Ross 1955	M F, NS, 12	Stewed apples, cooked in a Cu vessel, were served to 87 adults and 458 children; 3 adults and 9 children developed symptoms	Bitter taste, nausea, vomiting, and diarrhea
Roberts 1956	M, 24 yr, 1	≈400 g of CuSO$_4$ in water over a 4-mo period (≈2 g Cu per day)	Abdominal pain, hemolytic anemia
Sanghvi et al. 1957	M, 18, 20 yr, 2	Ingestion of CuSO$_4$ (amounts unknown) with suicidal intent	Sulfhemoglobinemia, acute renal tubular necrosis, azotemia, anuria, coma, and death
Wyllie 1957	F, NR, 15	Contaminated alcohol lemon cocktails from cocktail shakers containing Cu; estimated Cu ingestion, 5-32 mg	Weakness, abdominal cramps, headache, nausea, dizziness, vomiting in 10/13 nurses; possible causes, other than Cu, not excluded
Semple et al. 1960	M, NR, 18	Contaminated tea (≈250 mL) containing Cu at 44 mg/L; tea made with water from a Cu-lined, gas-heated boiler	Abdominal pain, vomiting, diarrhea, dizziness, and headache developed in 18 of 150 workers within 10 min after drinking the tea
Le Van and Perry 1961	M, NR, NR	Unspecified number of men on ship drank Cu-contaminated softdrink (Cu content not stated) from a vending machine	Several men developed mild nausea
Chowdhury et al. 1961	M F, adults, 20	12 women and 8 men attempted suicide by ingestion of unknown quantities of CuSO$_4$	Epigastric pain, nausea, jaundice, hemolysis, hemoglobinuria, proteinuria, glucosuria, hypotension, tachycardia, stupor; no deaths

TABLE 5-1 (*Continued*)

Reference	Sex, Age, Number	Exposures	Clinical Sequelae and Comments
Ghosh and Aggarwal 1962	M F, children, 32	31 children accidentally ingested unknown amounts of $CuSO_4$; one boy (age 11) given $CuSO_4$ for attempted homicide	Vomiting was chief complaint; the victim of attempted homicide developed severe hemolysis
Bohré et al. 1965	M, adults, 7	Naval officers consumed coffee prepared with water from a Cu-contaminated electrically heated boiler	Nausea, vomiting, malaise, collapse, no deaths
Chuttani et al. 1965	M F, 14-60 yr, 53	Oral ingestion with suicidal intent; estimated 1-30 g of $CuSO_4$	Diarrhea (29%), hemoglobinuria and hematuria (29%), anuria (27%), jaundice (23%), hypotension (8%), coma (8%), death (15%)
Fairbanks 1967	F, 22 yr, 1	Accidental ingestion of unknown amount of $CuSO_4$ solution	Nausea, vomiting, diarrhea, abdominal pain, icterus, melena, hemoglobinuria, proteinuria, mild anemia
Wahal et al. 1965	M F, 15-58 yr, 100	100 patients ingested $CuSO_4$ (dose unknown, mostly suicidal)	Hepatotoxicity (36%), renal toxicity (18%), cardiovascular toxicity (10%), neurotoxicity (10%), GI hemorrhage (6%), death (10%)
Nicholas 1968	M, NR, 20	Contaminated tea ($\approx$250 mL) containing Cu at >30 mg/L; tea made with water from a Cu-lined boiler	Diarrhea, nausea, vomiting
Singh and Singh 1968	M F, 13-85 yr, 40	27 males and 13 females; accidental or suicidal ingestion of copper salts	Variable hypercupremia, transaminasemia, hyperbilirubinemia, and renal insufficiency; 4 deaths
Salmon and Wright 1971	M, 1.25 hr, 1	Oral ingestion of drinking water (Cu at 0.35-0.79 mg/L) over a 3-mo period	Prostration, vomiting, red extremities, hypotonia, photophobia, peripheral edema, hypercupremia

Reference	Subject	Exposure	Effects
McMullen 1971	M, NR, ≥10	Oral ingestion of Cu-contaminated orange- or lime-flavored soft drinks, dispensed through Cu-containing bottle pourers; drinks contained Cu at 190 and 220 mg/L.	At least 10 members of a sports club vomited immediately after consuming the soft drinks
Mittal 1972	M, 22 yr, 1	Oral ingestion of ≈175 g of $CuSO_4$ (≈70 g Cu)	Severe abdominal pain, vomiting, diarrhea starting 1.5 hr post-ingestion; renal damage with hemoglobinuria; eventual recovery
Chugh et al. 1975	M, 27 yr, 1	Oral ingestion of ≥50 g of $CuSO_4$ (≈20 g Cu)	Cyanosis, oliguria, hemolysis, methemoglobinemia, death 16 hr post-ingestion
Agarwal et al. 1975	F, 41 yr, 1	Oral ingestion of $CuSO_4$ with suicidal intent; patient treated by hemodialysis at 13 hr after ingestion	Vomiting, diarrhea, hepatorenal failure, coma; hemodialysis was ineffectual; death; autopsy showed Cu in brain, heart, liver, kidney
Stein et al. 1976	F, 44 yr, 1	After the patient had ingested ethanol and diazepam, $CuSO_4$ (796 mg Cu) was administered as an emetic	Respiratory collapse, massive GI hemorrhage, hemolytic anemia, renal and hepatic failure, death; autopsy showed renal tubular necrosis and increased hepatic Cu content
Chugh et al. 1977	M F, 18-35 yr, 11	Report of 11 cases of acute renal failure in a series of 29 patients with acute $CuSO_4$ poisoning with suicidal intent.	Vomiting (100%), epigastric pain (45%), diarrhea (45%), oliguria or anuria (91%), jaundice (91%), coma (18%); death (45%); biopsy or autopsy showed renal tubular necrosis in 7 of 8 patients studied
Walsh et al. 1977	M, 1.5 yr, 1	Oral ingestion of 3 g of $CuSO_4$ (≈1.2 g Cu)	Hemolytic anemia was observed 2 days post-ingestion, with hematuria, glucosuria, proteinuria, cylindruria, hypercupremia, hypercupriuria; recovery period was ≈1 yr
Stenhammar 1979	?, 1-2.5 yr, 3	Oral ingestion for unknown duration of tap water containing Cu at 0.22-1 mg/L	Prolonged diarrhea with weight loss; symptoms disappeared with change of water source

TABLE 5-1 *(Continued)*

Reference	Sex, Age, Number	Exposures	Clinical Sequelae and Comments
Berg and Lundh 1981	M F, <3 yr, NR	Children at 7 Swedish kindergartens with Cu content in first-draw water from 0.35 to 6.5 mg/L	Association noted between Cu content of drinking water and diarrhea, but other possible causes of diarrhea were not studied
Spitalny et al. 1984	M F, 5, 7, 32 yr, 3	Oral ingestion of drinking water (Cu at 2-8 mg/L; median 3 mg/L) for 1.5 yr by a family living at the end of a copper water main	Episodic nausea, vomiting, abdominal pain occurred 5-20 min after drinking tap water in morning in 3 of 4 family members; the symptoms ceased with change of water source
Jantsch et al. 1985	M, 42 yr, 1	Ingestion of ≈250 g of $CuSO_4$ (≈100 g Cu) in attempted suicide	Protracted vomiting, hepatic failure, response to chelation therapy, eventual recovery
Müller-Höcker et al. 1988	M F, 1 yr, 2	Two infant siblings consumed for >9 mo Cu-containing tap water (Cu at 2.2-3.4 mg/L); the Cu was derived from Cu pipes	Micronodular cirrhosis with hepatic Cu storage, hepatosplenomegaly, jaundice, and hypertransaminasemia; 1 death
Knobeloch et al. 1994	M F, NR	Five studies of GI upsets associated with ingestion of Cu-containing drinking water	Higher incidence of symptoms following ingestion of first-draw water compared with flushed water

NR, not recorded.

difficult to control for potential confounding factors, such as microbial agents and their toxins. Following acute ingestion of copper salts (e.g., copper sulfate) in amounts that exceed approximately 1 g, systemic effects are generally observed. The effects include GI mucosal ulcerations and bleeding, acute hemolysis and hemoglobinuria, hepatic necrosis with jaundice, nephropathy with azotemia and oliguria, cardiotoxicity with hypotension, tachycardia and tachypnea, and central-nervous-system (CNS) manifestations, including dizziness, headache, convulsions, lethargy, stupor, and coma.

The systemic sequelae of acute ingestion of copper salts are highly variable. Oral intake of ionic copper usually induces immediate emesis, which reduces the quantity of residual copper available for absorption from the GI tract. Oral intake of copper that is bound to particulates in water or to proteins, lipids, and other constituents of foods is less likely to cause emesis, since the bound forms of copper show reduced bioavailability compared with ionic copper. The principal targets for acute copper toxicosis are the GI, hepatic, renal, hematological, cardiovascular, and CNS systems. There are few, if any, reports of musculoskeletal, dermal, ocular, immunological, carcinogenic, reproductive, or developmental effects in humans following oral ingestion of copper salts, even at high exposure concentrations.

Acute copper toxicosis, manifested by hemolysis, headache, febrile reactions, prostration, and GI symptoms, was observed in one child after a solution containing copper sulfate was applied to burned skin during a debridement procedure (Holtzman et al. 1966) and in numerous patients after inadvertent introduction of copper into the circulating blood during hemodialysis (Manzler and Schreiner 1970; CDC 1974; Lyle et al. 1976). In hemodialysis patients, ionic copper can be released from semipermeable membranes fabricated with copper or from copper tubing or heating coils of the dialysis equipment, especially when the dialysate has become acidic (Barbour et al. 1971; Blomfield et al. 1969, 1971; Klein et al. 1972). In two hemodialysis patients, copper intoxication was characterized by marked hemolysis, acidosis, methemoglobinemia, hypoglycemia, vomiting, epigastric pain, diarrhea, and headache, with fatal outcome (Matter et al. 1969). In one report, copper stopcocks in circuits used for exchange transfusions were identified as the source of potentially hazardous infusions of copper in neonates (Blomfield 1969). Administration of copper sulfate as an emetic was identified as another iatrogenic cause of acute copper toxicosis (Holtzman and Haslam 1968).

Wyllie (1957) describe one episode of acute GI symptoms associated with a presumed exposure to copper as a result of mixing alcoholic drinks in a copper-contaminated cocktail shaker. In reconstructing the exposure, the author concluded that the lowest adverse effect level was approximately 5.3 mg in ¾ fluid ounces or 10.65 mg of copper. However, significant questions have been raised about the suitability of those data for estimating toxic doses of copper (Donohue 1997).

Hopper and Adams (1958) presented five instances where faulty check valves in vending machines were responsible for carbon dioxide back flow and subsequent build-up of copper in vending machine water lines. The first drink in the morning can have a metallic taste, and cause salivation, nausea, vomiting, epigastric burning, or diarrhea (Hopper and Adams 1958).

Semple et al. (1960) reported an outbreak of copper poisoning from ingestion of tea that was contaminated with copper sulfate scale deposited in the water used to make the tea. The authors estimated that the total copper in the suspension was 44 mg/L. That estimate is unreliable, however, because exposure likely occurred after a large portion of the scale was dislodged in the vessel, and the water used to make the tea was not available for analysis.

The Centers for Disease Control and Prevention (CDC) reported multiple outbreaks of copper poisoning from ingestion of contaminated beverages (CDC 1974, 1975, 1977, 1996). In most cases, the copper concentration associated with illness was in excess of 30 mg/L. A major incident occurred in 1993-1994, where 43 individuals became ill from a single point source in a hotel. Exposures were estimated to range from 4.0 to 70 mg/L.

Recurrent GI illness, including nausea and vomiting, occurred in a Vermont family. Exposure was traced to a build-up of copper in the water overnight. Copper concentrations reached 7.8 mg/L (with a range of 2.8 to 7.8 mg/L) (Spitalny et al. 1984). Family members had increased copper concentrations in hair, but not blood. Relief of symptoms occurred when their drinking water was replaced by bottled water.

Knobeloch et al. (1994) investigated five individuals who ingested water above EPA's MCLG of 1.3 mg/L and reported abdominal symptoms. The authors suggested that increased copper in tap water can be a relatively common cause of GI symptoms.

Using survey methods to gather data on copper-induced illness from soft drinks, Low et al. (1996) queried 2,100 state and local departments of health and agriculture and water utilities. Although the response rate was only 40%, they found 70 incidents of copper poisonings affecting 462 people. Copper concentration data were available for 24 cases and were less than 10 mg/L in 6 of those 24 cases (range of 3.5 to <10 mg/L). Dose estimation from those data is difficult however, because little information is available on the amount of beverage consumed and on whether the drink was consumed with food.

Larger epidemiological studies have also investigated the relationship between exposure to excess copper environmentally and occupationally and adverse health effects. Recently, Buchanan et al. (1998) studied households from lists supplied by the Nebraska Department of Health. Households were chosen on the basis of their drinking-water copper concentrations, which were above 3 mg/L in 60 households, between 2 and 3 mg/L in 60 households, and below 1.3 mg/L in 62 households. A telephone interview was conducted with one adult from each household, addressing the occurrence of GI illness. A nested study was also performed where in-person interviews were conducted and copper water concentrations were measured. The investigators found no association between copper concentration and GI illness. The copper water concentrations ranged from 0.06 to about 5 mg of copper/L in the first draw water.

Roberts et al. (1996) examined communities made up of more than 100 people in Delaware. They considered communities in which 10% of the water samples, measured during a statewide survey in 1995, had copper concentrations greater than 5 mg/L. Four communities that met that criteria and one trailer park with older homes and acidic water were studied. First morning tap water was collected, and the households with concentrations in excess of 5 mg/L were revisited for study. Residents were interviewed once per week for 12 weeks and asked about GI symptoms. Although people with high concentrations of copper in their drinking water were slightly more likely to report becoming ill at some point during the study, there was no significant association of tap-water copper concentration and GI symptoms.

The committee had difficulty in determining how to use the available epidemiological data. Although copper concentrations were used to stratify exposures in communities, it is difficult to link copper concentrations to individual exposures. Not everyone in a household can drink first-draw water; therefore, the high exposure would be encountered by one person in a household. The epidemiological data provide some assurance that copper concentrations in first-draw water above the current MCLG do not produce a high frequency of adverse GI effects within a community. However, the data do not allow the conclusion that water consumed at those concentrations would not give rise to GI-related symptoms in individuals. As a consequence, the committee relied on experimental studies that involved administration of water containing identified concentrations of copper.

In one of the few experimental studies on copper in drinking water, Pizarro et al. (1999) designed a study to determine a threshold concentration for acute GI effects from copper in tap water. Sixty healthy adult women of low socioeconomic status from Santiago, Chile were randomly assigned to groups receiving 1, 3, or 5 mg of copper/L for 2 weeks, followed by 1 week of standard tap water. The women prepared the test water every morning and recorded their water intake and GI symptoms. Daily does of copper were not reported. Nausea, vomiting, abdominal pain or cramps, diarrhea, and food intolerance were recorded. Of the 60 participants, one woman recorded nausea only and two abdominal pain with 0 mg of copper/L of water. With 1 mg of copper/L, one woman recorded abdominal pain only, two recorded diarrhea only, one recorded diarrhea and vomiting, and one recorded all three symptoms. With 3 mg of copper/L, six women recorded nausea only, three recorded abdominal pain only, and one recorded vomiting only. Four women experienced diarrhea and abdominal pain. At 5 mg of copper/L, nausea only, abdominal pain only, diarrhea only, and vomiting only were recorded by five, two, three, and two women, respectively. One experienced both diarrhea and vomiting at 5 mg of copper/L. Therefore, a total of 21 subjects reported GI problems at some time during the study. Nausea, abdominal pain, or vomiting occurred 5%, 2%, 17%, and 15% of the time at 0, 1, 3 and 5 mg of copper/L, respectively. The data suggest that at or greater than 3 mg of copper/L can be associated with GI effects. The data also indicate a range of sensitivity in the population: 17% of the subjects reported symptoms at 3 mg of copper/L, while 85% did not report symptoms at 5 mg/L.

In follow-up to Pizarro et al. (1999), a larger experimental study involving controlled randomized trials (sponsored by the International Copper Association) is being conducted to determine the dose-response relationship more precisely for the GI effects of copper in drinking water (ICA, unpublished material, Oct. 13 1999). The results of that study are nearing completion as this report goes to press. The experimental protocol, methods, and individual data were made available to the committee. The study involved 60 adult volunteers in Ireland (Colerain), Chile (Santiago), and North Dakota (Grand Forks). Each subject drank 200 mL of water containing 0, 2, 4, 6, or 8 mg of copper/L. Symptoms were noted at various times up to 60 min at each testing center, and a 24-hr follow up was made for any other symptoms. Each subject received each of the doses in a randomized order once a week.

A dose-response for nausea was noted, although a lack of masking for taste might have affected the relationship. Not all individuals noted nausea even at higher doses, and a large variation in sensitivity among subjects is apparent. The fractions of those reporting nausea out of 180 subjects were 8, 7, 11, 25, and 44 at 0, 2, 4, 6, and 8 mg of copper/L of solution, respectively. Those results appear to be consistent with Pizarro et al. (1999), although they were considered too preliminary for further conclusions.

An additional study is under way to investigate the acute effects of copper in drinking water. No data are available from that study (L. Chaffin, League of Nebraska Municipalities, personal commun., Feb. 9, 2000).

In summary, there are inconsistencies among the data and they suffer from several limitations. There is little information on possible confounders and biases in the studies (including microbial water contamination and water averting behavior of residents), and there is an uneven distribution of risk in multi-member families. Copper concentrations often are not reliable because of a wide variability in first-draw concentrations and samples typically were not collected during the actual study period. Possible acclimatization is also not typically considered. In the experimental and epidemiological studies, sample sizes are small; therefore, the committee is concerned that effects on a sensitive population could have been missed. In addition, little work has addressed the health effects associated with exposure of infants and children to increased concentrations of copper in water. The only experimental study published in the literature indicates that GI symptoms arise from exposure at approximately 3 mg of copper/L (Pizarro et al. 1999).

It is important to stress the point that all of the above cases represent acute toxicity as a consequence of the consumption of water or beverages that contain high levels of copper.

For systemic effects, doses of copper in water associated with health effects differ from toxic doses in environmental media because of differences in bioavailability among water, food, and other environmental media. Copper ions in water have the highest bioavailability. The bioavailability of copper in the diet is a function of its solubility and also the types of complexes in which it is present. For example, complexes of copper with some amino acids and organic acids result in bioavailability similar to that of soluble copper sulfate (Wapnir 1998), whereas other dietary elements and certain amino acids can inhibit copper absorption.

CHRONIC TOXICITY

A major target of chronic copper toxicity is the liver. Liver toxicity is usually seen in specific populations, such as individuals with Wilson disease and children with various cirrhosis syndromes (see Chapter 4 for descriptions). However, there has been a case report of chronic ingestion of a high-dose copper supplement (30 mg per day for 2 years followed by 60 mg per day for 1 year) resulting in liver disease (O'Donohue et al. 1993). In that case, the pathological picture is similar to that seen in Wilson disease or the various childhood cirrhosis syndromes associated with excess copper exposure. Experimental animal studies have also demonstrated that ingestion of high amounts of copper in feed can lead to hepatic and renal disease (see section in this chapter on animal studies).

A paper by Scheinberg and Sternlieb (1996) is frequently cited as evidence that high concentrations of copper do not cause liver toxicity. That article reported on a study of deaths from cirrhosis among children under

6 years of age[1] exposed to tap water containing copper at 8.5 to 8.8 mg/L in three towns in Massachusetts. The authors reported that no deaths from cirrhosis occurred among those children. However, the crudeness of the end point measured (death), the potentially small number of children actually at risk, and the variability in exposure depending upon whether the child was breast fed or formula fed puts that conclusion into question.

The CNS can also be a target of chronic copper toxicity. Generally speaking, reports of neurotoxicity from chronic copper exposure have been limited to humans with Wilson disease. The CNS effects of copper in Wilson individuals are discussed in Chapter 4. Typically, neurological abnormalities have only been reported in animals administered very high doses of copper. Genetic animal models, such as the LEC rat (Mori et al. 1994; Kitaura et al. 1999), and Bedlington terriers with canine copper toxicosis do not have an increased susceptibility to the neurotoxic effects of copper (Owen et al. 1980, Hultgren et al. 1986).

Hemolytic anemia due to high concentrations of circulating copper can also occur. Anemia has been seen occasionally in Wilson-disease patients (Scheinberg and Sternlieb 1984; Brewer and Yuzbasiyan-Gurkan 1992) and with copper poisoning in sheep (Gooneratne et al. 1981). In both situations, the anemia is due to a concomitant acute hepatic necrosis (Scheinberg and Sternlieb 1984; Brewer and Yuzbasiyan-Gurkan 1992). That breakdown of the liver cells releases a very large amount of copper into the circulation, damaging red blood cells and causing the acute hemolytic anemia (Scheinberg and Sternlieb 1984; Brewer and Yuzbasiyan-Gurkan 1992). Cessation of menstruation, an increased incidence of gall stones and renal stones, a form of osteoarthritis, and some kidney-function abnormalities can also occur in acute, untreated Wilson patients (Scheinberg and Sternlieb, 1984; Brewer and Yuzbasiyan-Gurkan, 1992). After exposure to exogenous copper in large amounts, acute renal failure can occur and result in permanent renal damage (Chugh et al. 1975; Holtzman et al. 1966).

Reproductive and Developmental Toxicity

Small amounts of copper from intrauterine devices can prevent embryogenesis by blocking implantation and blastocyst development (Hurley and Keen 1979; Keen 1996; Hanna et al. 1997). In women with untreated Wilson disease, pregnancy is rare and often ends in spontaneous abortion.

---

[1] A total of 64,124 child years based on the average 0-5-year-old population from 1969 to 1991 multiplied by the 23 years in that period.

There are no reports in the literature of teratogenic effects associated with excess copper in humans. In a number of cases in which pregnancy continued despite the presence of a copper-containing intrauterine device, teratogenic effects of copper were not noted (Barash et al. 1990). In addition, there are no reports of abnormalities in the offspring of untreated Wilson patients.

The reproductive and developmental effects of excess copper in animals are discussed later in this chapter.

Genotoxicity, Mutagenicity, and Carcinogenicity of Copper

There have been numerous epidemiological studies looking at the relationship between exposure to copper, copper intake, serum copper, and various cancers. Those studies are complicated by the fact that serum copper concentrations can be increased as a result of cancer. Ceruloplasmin increases at times of stress (i.e., it is an acute-phase reactant) and 90% of serum copper is associated with ceruloplasmin. Wilson patients do have an increased risk of liver cancer, but only secondary to the liver cirrhosis associated with Wilson disease (Scheinberg and Sternlieb, 1984).

In a large prospective cohort study (more than 10,000 Dutch individuals), Kok et al. (1988) found no association between serum copper concentrations and overall cancer deaths. However, when the group was stratified by copper concentration, there was a significant increase in risk (odds ratio (OR) of 3.7; confidence limits, 1.5-9.1) for the subjects with the highest serum copper concentrations. Cetlnkaya et al. (1988) conducted a relatively small study comparing mean copper and zinc concentrations in 20 healthy women and 100 women with gynecological malignancies. They reported that mean serum copper concentrations were highest in the patients with malignant tumors. It is important to note that the serum concentrations of copper were measured after diagnosis and might have been due to increased ceruloplasmin. In addition, the analysis did not control for other possible confounders. Coates et al. (1989) used a case-control design among 5,000 Washington state employees and compared serum copper concentrations in 133 cancer cases and 241 controls among the employees. When the data were adjusted for potential confounders, there was no significant risk.

A case-control study by Cavallo et al. (1991) compared the mean dietary intake of copper in controls (noncancer patients) and patients diagnosed with primary breast cancer. There was no significant association between dietary intake of the copper and copper blood concentrations. Further, there was no significant association between copper intake and breast-cancer occurrence. In a subset of cases and controls in this study, there was a significantly increased serum copper concentration in cases compared with controls; however, there was no evidence of any association with serum copper concentration when they were examined by quintiles. Prasad et al. (1992) conducted a case-control study of 35 esophageal cancer patients and 35 controls in India and found no difference in serum copper concentrations between the cases and the controls.

Overvad et al.(1993) studied plasma copper concentrations in women who developed breast cancer between 1968 and 1985. In this prospective study, copper concentrations in 46 women with breast cancer were compared with a stratified random sample of 38 women. The adjusted odds ratios did not suggest an increased risk for breast cancer due to copper exposure, although when copper levels were stratified, the highest exposure group did have a significantly increased relative risk.

Dabek et al. (1992) studied 13 premenopausal and postmenopausal breast-cancer cases and compared them with 14 premenopausal and 11 postmenopausal noncancer controls. This prospective study was conducted to determine whether serum copper concentrations are altered significantly over a period of 1 year. Dabek reported a significantly higher serum copper concentration in the premenopausal breast cancer patients, but not in the postmenopausal patients, compared with the controls. In fact, postmenopausal breast cancer patients had significantly lower ceruloplasmin concentrations compared with their controls. Their lower concentrations might be due to lower estrogen concentrations. Dabek concluded that the copper-to-ceruloplasmin ratio in breast-cancer patients might reflect disordered copper metabolism in the disease.

Occupational studies have been difficult to interpret, because most exposures involve mixtures of metals. However, Logue et al. (1982) found no evidence of any increase in all cancers in 3,550 men who were employed in the tank house of copper refineries. No other studies have adequately addressed the question of copper carcinogenicity in the workplace.

In general, reviews of the epidemiological literature have highlighted the fact that most studies have determined serum copper concentrations only after diagnosis. Therefore, increased serum copper concentrations might be an effect of the cancer rather than a cause. Proper analysis of confounding variables is extremely rare, and any meaningful dose-response analysis is essentially absent from most studies. However, the few prospective studies that have been done provide little evidence for an association between copper and malignant disease. Therefore, there is inadequate evidence that copper plays any direct role in the development of cancer in humans.

Copper generates oxygen radicals via a Fenton-type reaction (Goldstein and Czapski 1986), and many investigators have hypothesized that excess copper might cause cellular injury via an oxidative pathway, giving rise to enhanced lipid peroxidation, thiol oxidation, and, ultimately, DNA damage. Such damage can occur when copper is used in combination with lipids, hydroquinone, and aliphatic and aromatic aldehydes (Li and Trush 1993; Becker et al. 1996; Glass and Stark 1997). Therefore, copper could enhance endogenous oxidative reactions that cause DNA damage.

Von Rosen (1964) reported that copper induces chromosomal breakage in plant cells in vitro. Other investigators have confirmed that copper (primarily tested as copper sulfate) is clastogenic in a number of different test systems, including the mouse bone-marrow system (Bhunya and Pati 1987). A dose-dependent increase in unscheduled DNA synthesis is also associated with in vitro exposure of rat hepatocytes to copper (Denizeau and Marion 1989). Most researchers report the chromosomal damage to be dose dependent. In contrast, copper sulfate does not induce DNA damage in prokaryotes (Matsui 1980), and copper chloride does not induce DNA strand breaks in phage PM2 in vitro (Becker et al. 1996).

Results from mutagenic studies are conflicting. Copper (II) sulfate is negative in the Ames *Salmonella* reversion assay (Moriya et al. 1983; Marzin and Phi 1985), the SOS Chromotest with *E. coli*. PQ 37 (Olivier and Marzin 1987), and several other bacterial mutagenicity assays (Iyer and Szybalski 1958; Matsui 1980; Clark 1953). However, Hansen and Stefan (1984) tested copper sulfate in its pesticide form and reported it to be mutagenic. Copper chloride has been uniformly negative in bacterial mutagenicity tests (Wong 1988; Nishioka 1975; Kanematsu et al. 1980).

Sensitive Populations

Infants appear to be more sensitive to both low and excessive dietary copper intake than adults. With respect to excessive intakes, infants are sensitive to elevated copper in water for both exposure and physiological reasons. Infants fed formula reconstituted with tap water would consume a high amount of tap water, particularly on a per body weight basis. They also have higher absorption and reduced capacity to excrete copper at higher doses relative to older ages (see Chapter 2). Similar to infants, children may also be more sensitive to copper than adults, although there are less data available on children. Cases of liver cirrhosis in infants resulting from elevated copper in formula or milk, however, appear to also have a genetic basis (see below). Genetic defects in copper metabolism might confer sensitivity to excess copper exposure (also reviewed in Chapter 4).

Carriers of Genetic Defects in Copper Homeostasis

As previously discussed in the Chapter 4 section Heterozygotes for Wilson Disease, there is evidence that presumed heterozygote carriers of mutations in the Wilson gene can have subclinical abnormalities in copper metabolism at typical levels of dietary copper intake. Therefore, if ingestion of copper is substantially increased, heterozygotes might develop copper-induced liver disease. If current case-frequency estimates are correct, 1% of the U.S. population is at risk. If the frequency is severely underestimated, as much as 2% of the U.S. population is at risk (see calculations in the Chapter 4 section Heterozygotes for Wilson Disease).

In Chapter 4, Tyrolean infantile cirrhosis (TIC), Indian childhood cirrhosis (ICC), and idiopathic copper toxicosis (ICT) are reviewed. All three diseases occur in infants or children; are familial, with an autosomal recessive genetic pattern; and often are associated with a documented or inferred increased ingestion of copper. Based on an experimental reconstruction, the concentration of copper was 10-63 mg/L (Müller et al. 1996). Those data support the hypothesis that there is a genetic susceptibility in many populations to disease from increased ingestion of copper affecting infants and young children. Although the gene mutations that underlie those diseases remain unknown, at least three genes might be hypothesized.

The first hypothesis is that the canine copper toxicosis gene underlies the genetic susceptibility in the childhood cirrhosis syndromes. That gene, which is different from the Wilson (Dagenais et al. 1999), is present at high frequency in Bedlington terriers (Yuzbasiyan-Gurkan et al. 1997) and could be present in the human population and be associated with cirrhosis in young children exposed to increased concentrations of copper in milk and water. However, the canine toxicosis gene defect in homozygous form produces a severe, lifelong disease that is not dependent upon high levels of copper ingestion. In contrast, TIC, ICC, and ICT are restricted to early life, appear to require increased copper ingestion, and do not require treatment in survivors. Therefore, with the different clinical course, homozygocity for the defective canine toxicosis gene is unlikely to be involved in the childhood cirrhosis syndromes. Heterozygosity remains another possibility.

A second hypothesis is that heterozygosity for Wilson gene defect combined with abnormal copper handling and increased copper exposure (Brewer in press) results in an increased sensitivity to copper in early childhood and underlies TIC, ICC, and many of the ICT cases. Assuming that a common use of copper pots caused ubiquitous exposure, the number of cases in the TIC epidemic is compatible with the hypothesis that the affected children were carriers of the Wilson gene. The population in the Tyrolean area was approximately 45,000 individuals. At the peak of the epidemic, there were about 3.8 infant cases per year. The estimated frequency of heterozygotes for the Wilson gene defect, 1%, could easily account for the incidence of TIC based on the number of infants in a population of that size. Furthermore, there appeared to be a genetic component to the disease, as it occurred in some families but not in others, and the component appeared to be recessive, because in many families multiple, but not all, siblings were affected. That pattern of disease frequency would be expected if heterozygosity for the Wilson gene resulted in susceptibility to the disease and only infants fed milk that contained the high concentrations of copper. Therefore, sensitivity to increased copper concentrations would act as a dominant genetic trait that manifests only in infants because of their exposure to copper. The gradual disappearance of the disease might have been due to a decrease in the use of copper or brass containers for boiling. Similarly, in ICC, a high intake of copper from milk or water seems to affect certain infants or very young children in a familial fashion (Tanner et al. 1983). Eleven of 132 cases reported by Tanner et al. (1983) were associated with increased copper in water.

The ICT data again suggest that a defective gene is present that makes infants and children unusually susceptible to high copper intake. Although the majority of TIC and ICC cases appear to be associated with high concentrations of copper in milk, nearly all of the ICT cases associated with increased copper exposure appear to be associated with increased copper concentrations in water rather than in milk. Müller et al. (1998) reviewed 15 studies of ICT. All the patients had increased hepatic copper concentrations, even those in whom drinking-water copper concentrations were not increased, suggesting that other copper sources must also be involved in ICT.

A hypothesis of Horslen et al. (1994) is that the ICT cases can be categorized into two disease subgroups based on age of manifestation, clinical course of disease, ultrastructure of the liver, and copper exposure. With the first subgroup, ICT manifests before age 4 and is generally associated with increased copper exposure (as in ICC and TIC), whereas with the second subgroup, ICT appears after about age 4 without identifiable sources of copper excess.

Of the cases reviewed, some ICT cases indicate a genetic predisposition (i.e., effects in siblings and consanguineous parents). In others, that information is not available. Among those age 4 and older, none of the cases reviewed had evidence of excess copper exposure (DuBois et al. 1971; Lim and Choo 1979; Lefkowitch et al. 1982; Maggiore et al. 1987; Horslen et al. 1994; Ludwig et al. 1996). Of the 10 cases reviewed involving bottle-fed children under age 4, copper in tap water was moderately to highly increased (above 1.3 mg/L) in seven cases (two are siblings; see Table 6-3 for copper concentrations) (Müller-Hocker et al. 1987, 1988; Baker et al. 1995; Bent and Bohm 1995; Walker-Smith and Blomfield 1973; Adamson et al.

1992; Aljajeh et al. 1994). The other three cases involve a breast-fed infant, a previously bottle-fed 29-month-old child with copper at less than 0.1 mg/L of drinking water, and a 2-year-old who had consanguineous parents and who at 2 months of age received a formula prepared with tap water containing copper at 1.13 mg/L of water for 3-4 weeks.

It should be noted that other possible sources of copper were not always investigated, and the actual exposure concentration in water is uncertain. Thus, it is difficult to determine the exact concentration of copper in drinking water needed to precipitate disease in susceptible populations of children. Nevertheless, some of the cases might be explained by the number of individuals who were heterozygous for Wilson disease and thereby more sensitive to copper exposure.

A third hypothesis is that an undiscovered copper susceptibility gene, in combination with the high concentrations of copper, is responsible for the childhood cirrhosis epidemics. That gene could be either recessive or dominant and result in susceptibility to increased copper concentrations in infants. For example, genes involved in copper homeostasis have been recently discovered in yeast and animals. Many of those genes probably have homologous human genes. Mutations in such genes could underlie copper sensitivity.

Regardless of whether heterozygosity for the canine copper toxicosis, Wilson disease, or an undiscovered copper homeostasis gene underlies childhood cirrhosis, the risk of increased copper intake remains if a genetic susceptibility to copper is involved in these syndromes.

Glucose-6-Phosphate Dehydrogenase Deficiency (G6PD)

It has been hypothesized that individuals with G6PD deficiency are at increased risk to copper exposure, because, in vitro, G6PD-deficient red blood cells are more susceptible to hemolysis and damage from copper than non-G6PD-deficient cells (Moore and Calabrese 1980). However, in vivo, about 90% of copper is covalently bound to ceruloplasmin and is not likely to cause red-blood-cell toxicity. With the exception of suicide attempts with copper salts, there is little evidence that low molecular weight copper increases significantly as a consequence of high dietary copper intake. Indeed, even in Wilson patients who have increased copper concentrations, hemolysis does not occur in the absence of hepatic necrosis. Therefore, the relatively small change in free copper in plasma that might result from a change of copper concentrations in typical diet or water is not likely be sufficient to alter the survival of G6PD-deficient red blood cells. Thus, individuals with G6PD deficiency would not be expected to have increased sensitivity to high dietary copper intake.

ANIMAL STUDIES

In this section, data on the toxicological effects of excess copper in experimental animals and the mechanism of copper toxicosis are reviewed. The appropriateness of animal models for assessing human toxicity to excess copper is also addressed. It should be noted that typical daily intake of copper in adult nonoccupationally exposed humans ranges between 0.9 and 2.2 mg (11.9 and 67.2 µg of copper/kg per day for the 10th and 90th percentile for typical intake of copper). The major route is oral, and the variability in copper intake reflects differences in dietary habits and agricultural and food processing practices. Levels of intake can be exceeded, particularly in cases where the drinking water contains high copper concentrations. Those factors should be kept in mind when human studies are contrasted with experimental data in animal studies of chronic exposure to copper (see Table 5-2). Unless otherwise stated, doses are reported as milligrams of copper per kilogram per day regardless of the form of copper administered.

Toxicity in Animals

The literature on copper toxicity in animals is replete with studies in a variety of species. It is beyond the scope of this review to summarize those studies and the reader is referred to previous reviews (NRC 1977; ATSDR 1990; IPCS 1998; Barceloux 1999). Key studies that illustrate copper toxicosis are reviewed to provide background.

Hepatic and occasionally renal changes are the most common effects found in animals that are fed high concentrations of copper. However, sensitivity to copper toxicosis is highly species dependent. In general, poultry appear to resist chronic copper toxicosis better than most mammals (NRC 1977). Sheep display toxicity even with relatively low copper concentrations in the diet, especially when the content of dietary molybdenum is low (0.5 mg/kg). Repeated doses of 1.5-7.5 mg of copper/kg of body weight per day as copper(II) sulfate or copper(II) acetate are associated with progressive liver damage, hemolytic crisis and ultimately death (IPCS 1998). Conversely, high levels of molybdenum in the diet of sheep have been associated with copper deficiency due to the formation of thiomolybdates in the rumen, which are potent anticopper agents (Auza 1983; NRC 1977; Aaseth and Norseth 1986; Dick et al. 1975; Mason 1990).

In nonruminant mammalian species, such as rats, mice, rabbits, pigs and dogs, significant toxic effects of copper are associated with long-term ingestion of high doses of copper, well beyond those tolerated by humans. Increased mortality and growth retardation have been noted in rats subse-

TABLE 5-2 Select Copper Effect Levels Observed in Long-Term Animal Studies

Species	Copper Form	Effects	LOAEL or NOAEL (mg Cu/kg/day)[a]	Exposure Duration	Exposure Route	Reference
Mouse	Copper sulfate	Developmental malformations	53 (NOAEL) 80 (LOAEL)	30 days before mating; in females exposure continued until sacrifice on day 19	Oral (feed)	Lecyk 1980
Mouse	Copper sulfate	Reproductive	399, M (NOAEL) 537, F (NOAEL)	92 days	Oral (feed)	NTP 1993
Mouse	Copper sulfate	Survival, body weight, general toxicity	44, M; 126, F (NOAEL)	92 days	Oral (feed)	NTP 1993
		Forestomach hyperplasia	97 (NOAEL) 187, M (LOAEL) 267, F (LOAEL)			
		Hepatic	1,060 (NOAEL)			
		Renal	1,060 (NOAEL)			
Mouse	Copper gluconate	Reduction in life span	42.5 (LOAEL)	Throughout life span	Oral (drinking water)	Massie and Aiello 1984
Rat	Copper sulfate	Neoplastic	27, M (LOAEL) 40, F (LOAEL)	280-308 days	Oral (feed)	Harrison et al. 1954
Rat	Copper acetate	Increase in hepatic SGOT activity	7.9 (LOAEL)	90 day study	Oral (drinking water)	Epstein et al. 1982
Rat	Copper sulfate	Poor growth, hepatic hypertrophy, necrosis, inflammation	270 (LOAEL)	105 days	Oral (feed)	Haywood 1985; Haywood and Loughran 1985
Rat	Copper acetate	Body weight, organ weight, and bone size	130 (LOAEL)	126 days	Oral (feed)	Llewellyn et al. 1985

Species	Compound	Effect	Dose	Duration	Route	Reference
Rat	Copper sulfate	Survival, body weight, general toxicity	16 (NOAEL)	92 days	Oral (feed)	NTP 1993
		Forestomach hyperplasia	16 (NOAEL) 33 (LOAEL)			
		Hematological	55 (NOAEL) 113 (LOAEL)			
		Hepatic	16 (NOAEL) 33 (LOAEL)			
		Renal	16 (NOAEL)			
		Reproductive	66 (NOAEL)			
Mink	Copper sulfate	Reproductive	6 (NOAEL) 12 (LOAEL)	9 months before, and for 3 months after mating	Oral (feed)	Aulerich et al. 1982
Pig	Copper sulfate	Hematological (decreased hemoglobin and hematocrit)	8.8 (NOAEL) 14.6 (LOAEL)	54 days	Oral (feed)	Kline et al. 1971
		Other (decreased body weight)	8.8 (NOAEL) 14.6 (LOAEL)			
Pig	Copper carbonate	Hematological	36 (LOAEL)	48 days	Oral (feed)	Suttle and Mills 1966
	Copper hydroxide	Hepatic, increased SGOT activity	36 (LOAEL)			
Dog	Copper gluconate	Elevated SGPT	8.4 (NOAEL)	6-12 months	Oral (feed)	Shanaman 1972
Rabbit	Copper sulfate	Marked hepatic toxicity	10 (LOAEL)	Up to 400 days and over	Oral (drinking water)	Tachibana 1952

[a]Concentrations are expressed as mg of copper/kg of body weight per day regardless of the form of copper exposure.
Abbreviations: NOAEL, no-observed-adverse-effect level; LOAEL, lowest-observed-adverse-effect level; M, males; F, females; SGOT, serum glutamic oxaloacetate transaminase; SGPT, serum glutamic pyruvic transaminase.

quent to chronic ingestion of 27-300 mg of copper/kg of body weight (IPCS 1998; Boyden et al. 1938; Haywood 1985). Deaths have been commonly attributed to anorexia and extensive hepatic centrilobular necrosis. In mice exposed to copper guconate equivalent to 42.5 mg of copper/kg per day in the drinking water throughout life, a 12.8% reduction in maximal life span was noted (Massie and Aiello 1984).

The highest reported no-observed-adverse-effect level (NOAEL) in the rat following chronic exposure in feed is approximately 130 mg of copper/kg per day (administered as copper acetate) for 18 weeks (Llewellyn et al.1985). The lowest-observed-adverse-effect levels (LOAEL) reported in the rat, based on hepatic effects, is 7.9 mg of copper/kg per day in feed (administered as copper acetate) for 90 days (increased serum glutamic oxaloacetate transaminase (SGOT) activity; Epstein et al. 1982). Pigs and rats appear equally sensitive to hepatic injury, and mice appear least sensitive (ATSDR 1990), having NOAELS of 1,060 and 97 mg of copper/kg per day (13 weeks of exposure in feed) for hepatic and gastric effects, respectively. In rabbits, chronic exposure to copper in drinking water (10 mg of copper/kg of body weight) was associated with marked hepatic toxicity, whereas dogs exposed to copper gluconate equivalent to 8.4 mg of copper/kg of body weight in feed had no toxic effects (Shanaman 1972; Shanaman et al. 1972; see IPCS 1998). The wide range of LOAELs and NOAELs in response to copper exposure reflects the sensitivity of the different animals and strains to copper, the exposure paradigm (duration of exposure), and the test performed (see Table 5-2). The LOAEL (and by inference the NOAEL) is also likely affected by the route of oral exposure, because copper that is administered in drinking water and in the absence of other food is much more readily absorbed across the GI tract.

Hepatotoxicity is the most prominent and characteristic systemic effect following chronic exposure to copper. The hepatic lesions seen in copper overload appear to vary from species to species. In humans, marked mitochondrial abnormalities are seen in Wilson disease, and diet overloaded rats show nuclear destruction and various membrane abnormalities (Alt et al. 1990). Copper overload affects a variety of functions by cytosolic proteins, membranes, and subcellular organelles. High concentrations of copper can result in increased rates of lipid peroxidation as the intracellular concentrations of reactive oxygen species can be increased via the Haber-Weiss reaction (Lindquist 1968; Miller et al. 1990; Halliwell 1989). Protein kinase C (PKC) activation, secondary to copper-mediated reactive-oxygen species generation, has also been seen in copper-induced cell death (Mudassar et al. 1992). Alternatively, Csermely and colleagues (1988) suggested that copper might directly activate PKC. Impairment of bile secretion, the predominant route for copper elimination, also results in excessive lysosomal copper accumulation, and, in turn, decreased lysosomal membrane fluidity, increased lysosomal pH, breakdown of membranes, and the leakage of lysosomal enzymes, such as phosphatases, into the cytosol (Lindquist 1967; McNatt et al. 1971; Myers et al. 1993; Nieminen and Lemasters 1996).

In rats fed high copper in the diet as copper sulfate, serum glutamic pyruvic transaminase (SGPT) activity was increased after 1 week of exposure at 100 mg of copper/kg per day (intake not calculated) (Haywood and Comerford 1980). Livers from male rats fed a high copper diet (1500 mg of copper/kg feed; intake not provided) for 16 weeks showed an increase in the number and diversity of lysosomes; swelling of smooth endoplasmic reticulum, mitochondria, and canalicular microvilli; and fragmentation of rough endoplasmic reticulum. Nuclear degeneration occurred early, culminating in lysis (Fuentealba and Haywood 1988). In rats exposed to copper sulfate equivalent to 300 mg of copper/kg per day for 1 week, centrilobular necrosis was noted (Haywood 1985). Centrilobular necrosis and inflammatory foci were also noted in the rat after 2-3 weeks of exposure to copper sulfate equivalent to 40-250 mg of copper/kg per day (Haywood 1980; 1985; Haywood and Comerford 1980; Haywood and Loughran 1985; Haywood et al. 1985a,1985b; Rana and Kumar 1980).

An extensive study (Shanaman 1972; Shanaman et al. 1972; see IPCS 1998) was conducted on the effects of copper in the beagle dog. That study was not available to the committee for review. Copper administered as copper gluconate for 6-12 months in the diet (equivalent to 0-8.4 mg of copper/kg of body weight per day) did not show significant toxic effects. A reversible increase in SGPT activity was noted in 2 of the 12 experimental dogs at the highest dose of copper (8.4 mg of copper/kg per day), an effect that was considered insignificant by a task group (IPCS 1998). There were no apparent gross or microscopic pathological lesions or changes in organ weight associated with this exposure paradigm. The insensitivity of the dog to copper toxicosis relative to the rat might be related to the dog's ability to rely more on transcuprein and low-molecular-weight complexes and less on albumin and ceruloplasmin for transport of copper to cells (Montaser et al. 1992). Total serum copper concentrations in the dogs are one-third those in the rat, and plasma ceruloplasmin concentrations are 8-fold less in the dog than in the rat (Montaser et al. 1992). Another speculation that might account for the lessened sensitivity of dogs to copper is that they might have a particularly effective copper transporter.

Renal effects have also been observed in some studies. In rats fed diets containing 270-540 mg of copper/kg per day for 15 weeks (administered as copper sulfate), copper in the kidney rose to a plateau concentration in 4 weeks (Haywood 1980). Copper protein in the cells of the proximal tubules could be detected after 2 weeks (Haywood 1980; Haywood et al. 1985a, 1985b). Widespread sloughing of necrotic copper-containing tubule cells became marked after 5 weeks of exposure to copper sulfate equivalent to 270 mg of copper/kg per day but declined subsequently as regeneration occurred (Haywood 1980; Haywood et al. 1985a; Haywood et al. 1985b). In the group receiving copper at 540 mg/kg per day, the toxicosis was prolonged (Haywood 1980). However, no renal changes were noted in rats exposed to copper sulfate equivalent to 100 mg of copper/kg per day in the diet for 2 weeks (Haywood 1980).

A comprehensive study was conducted by the National Toxicology Program (NTP 1993) looking at the effects of cupric sulfate in drinking water (2 weeks) and feed (2 and 13 weeks) in $B6CF_1$ mice and F344/N rats. Based on water intake and weight measurements, mice or rats consumed equivalent to between 0 and 368 mg of copper/kg per day or 0 and 97 mg of copper/kg per day, respectively. Consumption by mice or rats in the feeding study was between 0 and 783 mg of copper/kg per day or 0 and 325 mg of copper/kg per day (as cupric sulfate pentahydrate), respectively. In a 13-week feeding study, rats were given copper at 0 to 141 mg of copper/kg per day and mice were given 0 to 1,061 mg of copper/kg per day. Hematology, clinical chemistry, urinalysis, reproductive toxicity, and histopathology were evaluated. The reproductive end points are discussed in the reproductive section later in this chapter.

In the 2-week drinking-water studies, water consumption in the three highest dose groups of both species was reduced by more than 65%. Some animals in the third highest dose group and all animals in the two highest dose groups died. The authors attributed the deaths to dehydration. The only gross or microscopic change specifically related to cupric sulfate toxicity was an increase in the size and number of cytoplasmic protein droplets in the epithelium of the renal proximal convoluted tubule in male rats consuming excess copper at 45 mg of copper/kg per day (NTP 1993).

In the 2-week feed studies, rats and mice in the two highest dose groups had reduced body-weight gains compared with controls. The reduction was attributed to decreased feed consumption. Hyperplasia with hyperkeratosis of the squamous epithelium on the limiting ridge of the forestomach was seen in rats and mice; the more severe lesions were in rats. Inflammation of the liver, periportal to midzonal in distribution, occurred in rats consuming excess copper at 179 mg of copper/kg per day. Depletion of hematopoietic cells in the rat was evident in the bone marrow and spleen. Kidneys of rats consuming excess copper at 92 mg of copper/kg per day had an increased number and size of protein droplets in the epithelial of the renal cortical tubules (NTP 1993). Comparison of toxicity following exposure in drinking water and feed is a difficult task given the differences in consumption of copper in the two exposure paradigms. The studies, nevertheless substantiate the insensitivity of both species to copper toxicosis when contrasted with human exposure paradigms.

In the 13-week feed study, there were no chemically related deaths in rats or mice, and no clinical signs of cupric sulfate toxicity were recorded. At 13 weeks, mean body weight was lower in copper-exposed animals than in controls for both rats and mice receiving excess copper at 64 mg of copper/kg per day and 188 mg of copper/kg per day, respectively. In the rat, hepatocellular damage was apparent, with increases in serum SGPT and sorbitol dehydrogenase activities, as well as increases in 5'-nucleotidase and bile salts (restricted to males). Renal tubule epithelial damage was suggested due to increases in urinary glucose, N-acetyl-ß-D-glucosaminidase (a lysosomal enzyme), and SGOT (a cytosolic enzyme). Development of a microcytic anemia (i.e., decreases in mean cell volume, hematocrit, and hemoglobin) was noted, and increases in reticulocyte numbers suggested a compensatory response to the anemia by the bone marrow. Rats in the highest dose groups (more than 33 mg of copper/kg per day) had hyperplasia and hyperkeratosis of the forestomach, inflammation of the liver, and increases in the number and size of protein droplets in the epithelial cytoplasm and the lumina of the proximal convoluted tubules (NTP 1993). Those droplets stained strongly positive for protein but were negative for iron, PAS, and acid-fast (lipofuscin) staining methods. Copper was present in most of the protein droplets. Transmission electron microscopy of the livers of rats revealed increases in the number of secondary lysosomes in hepatocytes in the periportal area (NTP 1993).

Mice in the 13-week feed study appeared to be much more resistant to the toxic effects of cupric sulfate than rats, and no effects on kidney function or histology were noted. There was a dose-related decrease in liver weights (NTP 1993). In mice receiving excess copper at 188 mg of copper/kg per day, there was a dose-related increase in hyperplasia, with hyperkeratosis of the squamous mucosa on the limiting ridge of the forestomach. Minimal positive staining for copper was present in the liver and was limited to mice given high doses (more than 766 mg of copper/kg per day).

Reproductive and Developmental Toxicity

An EC_{50} of 0.15 mg of copper/L (in the aqueous medium) has been reported to be teratogenic to frog embryos. The effects seen were defects of the eye, gut, notochord, and heart (Luo et al. 1993). The teratogenicity of excess copper in mammals has not been established. However, no defects are seen in newborn rats, hamsters, rabbits, sheep, or guinea pigs that were experimentally exposed to a uterine environment of high copper (Keen et al. 1982).

Keen et al. (1982) studied the effect of feeding pregnant mice diets high in copper under varying conditions. Mice fed diets containing up to 500 mg of copper/kg had normal litters. However, mice fed diets containing 2,000 mg of copper/kg during pregnancy did not carry their pregnancies to term. Mice exposed to that diet for only 5 days of pregnancy (days 7 to 12 of gestation) had resorption frequencies of more than 50%. The diet fed before and after the 5-day period contained 250 mg of copper/kg. Surviving fetuses were not visibly malformed, and their copper content was not appreciably higher than that of fetuses from dams fed diets containing 250 mg of copper/kg body weight throughout pregnancy. The high-copper diet caused a severe reduction in maternal food intake and a reduction of maternal body weight. It is thought that this caloric deprivation, rather than the excess of copper per se, was the cause of resorption. In pregnant mice, short periods of fasting (40 hr) can result in total litter resorption (Runner and Miller 1956). The importance of monitoring food intake in studies assessing the teratogenic potential of a nutrient is emphasized by the above results.

Ferm and Hanlon (1974) have reported that copper (10 mg of copper/kg) injected intraperitoneally (i.p.) on day 8 of gestation in the hamster is teratogenic. Fetal resorption, kinked-tail, thoracic and ventral hernias, microphthalmia, cleft lip, and ectopic cordis were among the abnormalities found. In rats, i.p. injection of copper from day 7 through day 10 of gestation resulted in a resorption frequency of 50% (Marois and Bovet 1972).

NTP (1993) examined vaginal cytology and sperm morphology, motility, and density, following a 13-week exposure to copper in feed, as previously described. No adverse effects on any of the reproductive characteristics measured in rats or mice of either sex were reported (NTP 1993). Reduced neonatal body and organ weights have been seen in the offspring of rats at doses of copper in excess of 30 mg of copper/kg of body weight per day over extended time periods; similarly, fetotoxic effects and malformations are seen with high doses (more than 80 mg of copper/kg of body weight per day) (IPCS 1998). In those cases, food intake was reduced, and that could account for the negative reproductive effects.

In summary, injected copper can be teratogenic. When administered in the diet, copper can be teratogenic if the amounts are high enough to cause marked inanition. Periods of starvation can be teratogenic or embryo lethal in many species.

Neurotoxicity

Generally speaking, neurotoxicity from copper seems to occur only in humans with Wilson disease; however, in various animal models in which copper ingestion is increased, some neurological effects have been observed. In rats maintained on a 10% casein diet, daily administration of manganese chloride (1 mg of magnesium/mL of drinking water) and copper sulfate (equivalent to 250 mg of copper/kg of diet; equivalent to about 20 mg of copper/kg of body weight per day) for 30 days resulted in learning and memory impairment (Murthy et al. 1981). The behavioral changes were associated with a marked accumulation of copper in the brain. Combined exposure to copper and manganese also produced increases in the dopamine (DA) and norepinephrine (NE), and a depression in 5-hydroxytryptamine (5-HT). In contrast, daily administration of 1,250 mg of copper sulfate/L of drinking water (0.125%; equivalent to about 46 mg of copper/kg of body weight per day) for a period of 11 months in weaning rats was not associated with changes in concentrations of dopamine in the brain (de Vries et al. 1986). However, the concentration of 3,4-dihydroxyphenylacetic acid in the corpus striatum was lowered (25% decrease) in the copper-exposed group. Saturation studies of the striatal D-2 dopamine receptors indicated that copper increased receptor affinity, with a trend for decreased receptor number (de Vries et al. 1986). In genetic animal models, such as the Long-Evans cinnamon rat (LEC), and in the canine copper toxicosis common in Bedlington terriers, no neurological abnormalities have been reported (Kodama 1996; Brewer et al. 1992).

Taken together, the chronic toxicity information indicates that, in the absence of genetic abnormalities, animals (with the exception of sheep) are not very sensitive to copper. Furthermore, given the massive doses (described above) required to induce chronic toxicity in mammalian animal species that have balanced mineral intake, animal studies provide little information on the mechanisms that underlie copper toxicity relevant to human dietary concentrations of copper. It is true that copper toxicity can occur in animals if they are given very high concentrations of copper in their food or water; however, interpretation of this finding must be bounded by the essentiality of copper at lower concentrations. The requirement for copper in tissues is under tight homeostatic control mechanisms (see Chapter 2). Cellular copper transport processes are required by all organisms for the optimal utilization of copper and the avoidance of toxicity due to excess copper. Chronic toxicity is, therefore, likely to occur when such control mechanisms are impaired or overcome by massive intakes of copper.

Carcinogenicity

Several animal studies have investigated the carcinogenic potential of various copper compounds. When judged against current methods, the data are not optimal, because most of the studies are dated and are single oral-dose studies (Tachibana 1952; Harrisson et al. 1954; Howell 1958;

Carlton and Price 1973). Few studies used parenteral exposure (Stoner et al. 1976). However, the data clearly do not provide any suggestion that copper is carcinogenic in animals. Data from the LEC rat, which has been used as a model for Wilson disease, support the hypothesis that the cirrhotic effects of copper can play a role in hepatic cancer.

The ability of metal ions to damage DNA and cause mutagenesis has also been analyzed with reversion and forward mutation assays using single-stranded DNA templates. Incubation of phi X174 am3 DNA with Fe^{2+} in vitro leads to mutagenesis when the treated DNA is transfected into *Escherichia coli* spheroplasts (Loeb et al., 1988). In the same assay, the frequency of mutants produced by copper ions was shown to be greater than that by Fe^{2+} (Tkeshelashvili et al., 1991). Although there are some data to suggest that oxidative DNA damage is important in the etiology of breast cancer (Malins et al. 1996), few human data are available to directly address that hypotheses.

Mechanisms and Animal Models for Copper Toxicity

Mechanism of Acute Copper Toxicity

Although the symptoms associated with acute and chronic copper toxicity have been well defined in humans, the mechanisms of copper toxicosis and the concentration at which toxicosis occurs remain poorly understood. A number of investigators have studied the mechanism of the gastric response to excess copper in an effort understand the implications of acute exposure to excess copper in drinking water. Wang and Borison (1951) noted that the latency for the gastric response is usually short (less than 1 hr) and that upper GI effects predominate. They severed the gastric nerves in dogs to determine if neuronal stimulation was necessary for the emetic response to copper exposure. In their work, when both the vagus nerve and sympathetic nerves were severed, the emetic dose increased approximately 8-10-fold, and the latency response rose by a factor of 10. Subsequent studies, most recently using the ferret as a model, suggest that stomach infusions are necessary to produce the emetic response (Makale and King 1992). Recent research also suggests that the emetic response can be blunted at the CNS level through the use of neurokinin-1 (NK-1) receptor antagonists (Saito et al. 1998). Therefore, the mechanism of the GI response to acute copper toxicity appears to be direct irritation of the stomach by copper ions. The emesis is primarily mediated neuronally but is affected by individual sensitivity, the volume of copper-containing material ingested, the state of the copper (free versus bound), and the presence or absence of gastric contents. There has also been some suggestion that adaptation might occur, making repeated exposure more tolerable.

Mechanism of Chronic Copper Toxicity

Genetic variants in animals provide a unique opportunity to study copper toxicosis. The LEC rat is an inbred mutant strain, isolated from Long-Evans rats, with spontaneous hepatitis. Approximately 40% of LEC rats die from fulminant hepatitis. The remaining 60% survive to develop chronic hepatitis and, subsequently, liver and kidney tumors (Mori et al. 1994; Kitaura et al. 1999). Therefore, the LEC rat has been used as an animal model for studying the role of chronic hepatitis in the development of liver cancer.

The LEC rat manifests increased hepatic copper, defective incorporation of copper into ceruloplasmin, and reduced biliary excretion of copper. Therefore, it serves as an important animal model for studying the etiology of Wilson disease (Li et al. 1991; Wu et al. 1994; Mori et al. 1994; Cuthbert 1995; Vulpe and Packman 1995; Harris and Gitlin 1996; Koizumi et al. 1998). Recent studies on the structure and expression of the ATP7B gene protein (ATP7B) support its role as a copper transporter involved in the intracellular trafficking of copper in hepatocytes. In LEC rats and Wilson disease, copper accumulation is associated with a defective copper-transporting P-type ATPase, resulting in reduced biliary excretion of copper (Muramatsu et al. 1998; Bingham et al. 1998; Terada et al. 1998; Terada et al. 1999). Wu et al. (1994) cloned cDNAs from the rat gene homologous to the human Wilson gene (ATP7B) and identified a partial deletion in that gene in the LEC rat. At least 900 bp of the coding region at the 3' end, including the crucial ATP binding domain, is missing from the gene (Wu et al. 1994).

Comparative dose-response relationships and hepatic or renal changes between the LEC rat and other rat strains are generally not available, because the LEC rat spontaneously develops tissue damage in the absence of copper fortified diets. Although LEC rats develop copper toxicosis after consumption of regular rodent diets (6 mg/kg),[2] it is necessary to fortify the diet or drinking water with massive amounts of copper for other strains to see copper toxicosis (see the above section Toxicity in Animals).

To investigate the role of the gene product (ATP7B) in hepatocytes, an expression plasmid carrying full-length complementary DNA for the human Wilson gene was constructed and expressed in hepatocytes of LEC rats (Nagano et al. 1998). ATP7B localized to the membrane with a molecular weight of 155 kilodaltons (kD). Upon expression of ATP7B in hepatocytes from LEC rats, the protein was present in the trans-Golgi network and at the plasma membrane, a distribution pattern similar to that of

---

[2]Typical Analysis of Diet #F3156 Rodent Diet, American Institute of Nutrition-93G.

Menkes-disease protein (ATP7A). Cotransfection and coexpression of the human Wilson gene and ceruloplasmin gene in cultured hepatocytes indicated that ATP7B always accompanied the distribution of ceruloplasmin at the perinuclear region but that part of ATP7B localized irrespective of the distribution of ceruloplasmin. Accordingly, ATP7B was proposed to localize to the trans-Golgi network and to transport copper into this compartment, ensuring optimal delivery of copper to the apo-ceruloplasmin (Nagano et al. 1998). In contrast, a fraction of ATP7B that was not accompanied by ceruloplasmin in the perinuclear region and at the plasma membrane was proposed to contribute to efflux of copper from the hepatocytes. The distribution patterns of ATP7B in hepatocytes might, therefore, explain the dual roles of this P-type ATPase in hepatocytes (Nagano et al. 1998).

Chronic copper toxicity in LEC rats is associated with the uptake of copper-loaded metallothionein (MT) into lysosomes, where it is incompletely degraded and polymerized to an insoluble material containing reactive copper (Koizumi et al. 1998). This copper, together with iron, has been postulated to catalyze Fenton-type reactions and lysosomal lipid peroxidation, leading to hepatocyte necrosis (Ma et al. 1997; Koizumi et al. 1998; Klein et al. 1998). Subsequent to phagocytosis by Kupffer cells, the reactive copper might amplify liver damage either directly or through stimulation of those cells (Klein et al. 1998). There is no general consensus regarding the role of MT in lipid peroxidation as the primary mechanism mediating copper toxicity. In vitro studies suggest that other thiol-rich cellular proteins might represent the primary site of copper-induced injury (Sokol et al. 1989). In that model, disturbances in the normal function of glutathione (GSH) in complexing copper soon after its uptake into the cell and the subsequent transfer of the complexed metal to MT where it is normally stored are cited as a potential mechanism of copper toxicosis (Freedman et al. 1989; Steinebach and Wolterbeek 1994).

Copper toxicosis in the LEC rat has been postulated to result from liberation of massive concentrations of copper from MT when the intracellular capacity of MT synthesis was bypassed (Suzuki 1995; Rui and Suzuki 1997). Copper containing MT exhibited antioxidant properties in the presence of zinc. When zinc was not present, however, MT liberated cuprous ions, exhibiting prooxidant activity (Suzuki et al. 1996). The removal of copper complexed to molybdenum into the bloodstream was postulated to depend upon the amount of copper accumulating in MT (Ogra and Suzuki 1998).

The fact that expression of heme oxygenase-1 (HO-1), an inducible isoform of heme oxygenase, and HO-2, a constitutive isoform of heme oxygenase, was enhanced in the LEC rat also supports that conclusion (Matsumoto et al. 1998). The high expression of HO-1 in the LEC rat was postulated to arise from the oxidative stress caused by the accumulation of free copper, free iron, and free heme concentrations, thus representing an adaptive response to oxidative stress. Depletion of hepatic selenium in the LEC rat and reduced capacity to protect cells from copper-induced free-radical damage have also been suggested as potential mechanisms for enhanced oxidative stress (Downey et al. 1998). Ma et al. (1997) speculated that generation of free radicals by copper precipitates the induction of hepatic DNA damage and oncogenesis.

Cellular copper homeostasis was recently studied in vitro in hepatic cell lines from the liver of LEC rats (Nakamura et al. 1995). Cells from the LEC rats accumulated larger amounts of copper than did control cells when the concentrations of copper in the culture medium exceeded 5 µM. However, the secretion of ceruloplasmin from the cultured cells was not reduced in hepatocytes from LEC cells compared with controls. Additional studies confirmed that the genetic defect in LEC rats did not alter the biosynthetic and secretory pathways of ceruloplasmin, and that the intracellular copper concentration did not regulate the synthesis and processing of ceruloplasmin in the cultured hepatocytes. Accordingly, the copper transporting ATPase encoded in the Wilson disease gene might not be an integral part of the biochemical mechanism of copper incorporation into apo-protein (Nakamura et al. 1995). However, given that Nakamura et al. (1995) used high concentrations of copper histidine, in which copper is in the cuprous form and can bypass normal transport mechanisms, the conclusion regarding the processing of ceruloplasmin in the LEC rat does not represent a relevant physiological situation.

The available data on the mechanisms of carcinogenicity of copper in the LEC rat are sparse. The DNA binding activity of the serum response factor (SRF) was reported to be higher in the liver of LEC rats (approximately 2-fold) than in that of Wistar rats (Maeda et al. 1997). There was a close correlation between the intensity of the activity and the concentrations of copper in the nuclear protein. The DNA binding activity of Sp1, on the other hand, showed similar levels in both LEC and Wistar rats. It has been postulated that SRF might play an important role in the development of hepatocellular carcinoma in LEC rats by mediating induction of the protooncogene c-*fos* and that the copper in nuclear protein might be involved in the activation of SRF (Maeda et al. 1997).

In addition, LEC rats exhibit abnormal expression of γ-aminobutyric acid$_A$ (GABA$_A$) receptor subunit genes in the brain, leading to an increase in GABAergic tone (Follesa et al. 1999). Neurochemical disturbances involved in abnormal catecholamine metabolism in the cerebral cortex of the LEC rat before excessive copper accumulation have also been described (Saito et al. 1996). It is unclear whether brain defects in the LEC rat are characteristic of human Wilson disease (Kodama 1996). The latter are caused primarily by neuronal damage due to copper deposition in the CNS, and at times they are caused by encephalopathy secondary to hepatic dysfunction. It is noteworthy that deposition of copper in the CNS of LEC rats has not been reported (Kodama 1996).

The molecular defect in ATPase of LEC rat alludes to its validity as a model of Wilson disease. Despite the abundance of clinical and biochemical similarities, however, the differences between the ATPase defect in LEC rats and Wilson disease must be recognized (see Table 5-3). In LEC rats, serum ceruloplasmin concentrations are almost normal, and the oxidase activity of ceruloplasmin, which represents holo-ceruloplasmin, is low in Wilson disease. Hepatocellular carcinoma occurs frequently in the LEC rat, and cirrhosis does not. It is unclear whether these difference are species specific (rat vs. human) or due to the primary defect (Kodama 1996). It is clear that the effects of dietary copper loads in rats that do not carry genetic defects in ATP7B do not mimic those of Wilson disease. Therefore, the LEC rat model offers a unique opportunity to study the etiology of this disease.

Toxic milk (tz), an autosomal recessive mutation in mice and a similar mutation in LEC rats in P-type ATPase, has also been described (Rauch, 1983). Mutant females produce offspring that exhibit poor growth, hypopigmentation, tremors, and ultimately die at 2 weeks of age. The mutants themselves accumulate large concentrations of copper in the liver and ultimately develop hepatic disease. Recent studies in the tx mouse have demonstrated that MT binds to copper in the liver and that high copper and MT concentrations are present. MT was postulated to enhance lipid peroxidation and account for genotoxicity and apoptosis (Deng et al., 1998).

A third inherited form of copper toxicosis has been characterized as Indian childhood cirrhosis (ICC). A homozygous defect in the ATP7B gene has been excluded as a cause of human copper overload (van de Sluis et al. 1999), but a 50-kD purified major copper-binding protein (MCuBP) was recently identified as a potential contributor to the total copper-binding activity in the ICC liver (Prasad et al. 1998). Because copper administration alone at relevant dietary concentrations has not been shown to induce cirrhosis in animals, synergy between copper and a second hepatotoxin has been suggested (Tanner and Mattocks 1987; Aston et al. 1998; Tanner 1998) in the etiology of ICC. The hypothesis that ICC results not only from copper overload but also a from a second hepatotoxin has been recently tested. Rats chronically fed with excess copper in combination with a pyrrolizidine alkaloid (PA), retrorsine, were evaluated for morphological damage in a number of tissues. Increased plasma bilirubin, falling plasma albumin, histological changes, and massive accumulation of copper in the liver occurred. When retrorsine was given alone to lactating rats, no adverse effects were seen, although the suckling newborns did show

TABLE 5-3 Comparison of LEC Rats with Wilson Disease

	Wilson Disease	LEC Rats
Inheritance	Autosomal recessive	Autosomal recessive
Copper concentration		
Liver	Severely high	Severely high
Biliary	Low	Low
Serum	Low	Low
Brain	High	Normal
Kidney	High	High
Urinary	High	High
Serum ceruloplasmin		
Oxidase activity	Low	Low
Amount of protein	Low	Almost normal
Serum SGOT, SGPT levels	Increase	Increase
Serum immunoglobulin G levels	High	Low
Hepatic pathology		Similar to Wilson disease
Outcome	Liver cirrhosis	Hepatocellular carcinoma
Therapy with chelators	Effective	Effective

Abbreviations: SGOT, serum glutamic oxaloacetate transaminase; SGPT, serum glutamic pyruvic transaminase.
Source: Modified from Kodama 1996.

liver accumulation of copper. When copper and retrorsine were subsequently administered, the rats developed severe liver damage with retention of copper. Retrorsine passing to rat neonates via breast milk led to an accumulation of hepatic copper and an impairment in the rise in serum ceruloplasmin, suggesting either a decline in synthesis of ceruloplasmin, or a failure of copper to incorporate into the apo-protein (Aston et al. 1996). Retrorsine also caused a decrease in hepatic MT and serum albumin levels (also indicative of decreased protein synthesis) and reduced hepatic DNA levels (indicative of decreased cell number but increased cell size). An accumulation of copper in liver, with a reduction in copper-binding proteins could result in an increase in the pool of copper which is available to generate reactive oxygen species. This could explain the synergistic hepatotoxicity of copper and retrorsine (Aston et al. 1996). While the above model produced cirrhosis, significant differences from the histology seen in the human disorder were noted (Aston et al. 1998). The lack of an animal model, the inconsistent relationship between liver copper concentrations and liver damage, and the rarity of liver disease in adults suggest that other factors contribute to the etiology of ICC. Therefore, the hypothesis that ICC results from copper and a second hepatotoxin requires additional testing.

Finally, a form of chronic copper toxicosis has been reported in certain Bedlington terriers (Owen et al. 1980). The condition is inherited and leads to progressive chronic hepatic degeneration (Hultgren et al. 1986). The DNA microsatellite marker C04107 has been linked to the copper toxicosis locus in Bedlington terriers, and it is used diagnostically to detect the disease allele (Yuzbasiyan-Gurkan et al. 1997; Holmes et al. 1998). The copper toxicosis locus in Bedlington terriers is located on dog chromosome 10 in a region syntenic to human chromosome region 2p13-p16 (van de Sluis et al. 1999; Dagenais et al 1999). Hepatic concentrations of copper in genetically afflicted Bedlington terriers are extremely high, and plasma ceruloplasmin concentrations are normal (Owen et al. 1980). Despite histological evidence of hepatitis in young dogs and cirrhosis in older ones, plasma levels of hemophilic factors VIII, IX and XI were observed to be above normal and were more closely related to the age of the dog than to hepatic copper concentrations (Owen et al. 1980). At the morphological level, ultrastructural and microanalytical techniques identified copper peaks in lysosomes, the nucleus, and the cytoplasm in descending order and profound cellular changes (Haywood et al. 1996). Hepatocytes generally appeared shrunk with compacted electron dense organelles; nuclei were contracted, misshapen with chromatin condensation and fragmentation; and apoptotic bodies were identified in sinusoids. Excess copper in the Bedlington terriers was initially sequestered in lysosomes, but following increasing saturation of this compartment, nuclear copper accumulation and DNA damage occurred. Apoptosis followed, probably mediated by induction of p53 protein (Haywood et al. 1996). Hepatic copper concentrations in copper toxicosis-afflicted Bedlington terriers range from 1,000 to 10,000 µg/g of dry weight (Owen et al. 1980), compared with 200 to 3,000 µg/g of dry weight in patients with Wilson disease and below 50 µg/g of dry weight in normal humans. Furthermore, the plasma ceruloplasmin of genetically afflicted Bedlington terriers is not altered (Kodama 1996); blood ^{64}Cu concentrations 24 hr after administration of ^{64}Cu shows no difference between affected and unaffected Bedlington terriers (Brewer et al. 1992); and Kayser-Fleischer rings (characteristic of copper accumulation at the periphery of the cornea) and neurological dysfunction are always absent in afflicted Bedlington terriers. Thus, the Bedlington terrier model must have a new and as yet unidentified gene that is important for copper homeostasis.

CONCLUSIONS

• Acute GI effects of copper, including nausea and vomiting, have been seen in case reports and epidemiological studies. Dose-response information is difficult to determine from those studies.
• Recent controlled human experimental studies have demonstrated a dose-response relationship for the acute GI effects of copper.
• Acute copper toxicity does not seem to pose a significant reproductive risk for humans. In experimental animals, high concentrations of dietary copper do not pose a reproductive risk unless food intake is reduced. High concentrations of copper given by injection can be teratogenic, but the significance of that finding in humans is unclear.
• Copper metal is inactive in most assays of mutagenicity, although it can induce chromosomal and DNA damage via a free-radical-mediated mechanism under the appropriate conditions.
• There is inadequate evidence that copper plays a direct role in the development of cancer in humans.
• In sensitive human populations, the major target of chronic copper toxicity is the liver. In Wilson disease, neurological toxicity also occurs.
• Based on cases of TIC, ICC, and ICT, there appears to be a subset of the population sensitive to hepatic copper toxicity. Cases generally occur in infants or young children, have a familial pattern suggestive of recessive inheritance, and usually involve increased ingestion of copper in milk or water. The data suggest that a gene causes a predisposition to copper-induced liver cirrhosis. Given that some heterozygous carriers of the Wilson gene accumulate abnormal concentrations of copper, a reasonable hypothesis is that carriers of mutations of this gene comprise the infants with copper toxicosis disorders. An alternative hypothesis is that an unknown copper-susceptibility gene is present in many populations. Irrespective of which hypothesis is correct, increase in the ingestion of copper should be cautioned against until the hepatic susceptibility is clearly identified.
• In general, studies on the toxicity of copper in animals provide little information except for some data on physiological, biochemical, and pathological aspects of copper metabolism or chronic toxicity relevant to human dietary concentrations of copper.
• Although animal models provide some qualitative insight into the toxicology of copper, they are of limited value for establishing dose-response relationships in humans.
• The LEC rat, an inbred mutant strain isolated from the Long-Evans rat, is prone to copper toxicosis because of a defective ATP7B copper transporter. This illustrates how genetic errors in experimental animal models can result in syndromes that more closely represent the human situation. Therefore, such models are useful for studying human genetic defects.

- There are few studies in animals that evaluate copper in drinking water. Therefore, the differences in the bioavailability of copper in food versus drinking water are not well established.

RECOMMENDATIONS

- Although maternal exposure to high concentrations of dietary copper during pregnancy are not teratogenic, the potential developmental effects associated with exposure to high concentrations in the diet during the early postnatal period have not been well characterized. This area requires additional study.
- Epidemiological studies should be carried out to study the effects of long-term exposure to elevated copper in drinking water, as well as solid diet in sensitive populations. Hepatic toxicity should be a focus of such research.
- The frequency of the Wilson-disease gene defect should be established.
- The potential role of genetics that underlie infant and childhood copper toxicosis should be examined.
- Studies should be conducted in the LEC rat to determine the role of ATP7B in copper transport and the cellular and molecular mechanisms of tissue injury resulting from copper accumulation.
- LEC rats should be outbred with Long-Evans rats to create rats heterozygous for the ATP7B transporter. The interaction of genetic predisposition and copper overload in the new rat could then be evaluated.
- Additional studies should be conducted on specific mutations involved in copper transport. Additional genetic models of impaired copper metabolism need to be generated. The genetic models should also afford the possibility to test various pharmacological modalities for their potential to attenuate copper toxicosis.
- Animal studies on the effects of chronic exposure to copper in drinking water should be carried out to determine the differences in the bioavailability of copper in food versus drinking water.

REFERENCES

Aaseth, A. and T. Norseth. 1986. Copper. Pp. 233-254 in Handbook of Toxicology of Metals, 2nd Ed., L. Friberg, G.F. Nordberg and V. Vouk, eds. Amsterdam: Elsevier.

Adamson, M., B. Reiner, J.L. Olson, Z. Goodnam, L. Plotnick, I. Bernardini and W.A. Gahl. 1992. Indian childhood cirrhosis in an American child. Gastroenterology 102(5):1771-1777.

Agarwal, B.N., S.H. Bray, P. Bercz, R. Plotzker, E. Labovitz. 1975. Ineffectiveness of hemodialysis in copper sulfate poisoning. Nephron 15:74-77.

Aljajeh, A., S. Mughal, B. al-Tahou, T. Ajrawi, E.A. Ismail and N.C. Nayak. 1994. Indian childhood cirrhosis-like liver disease in an Arab child. A brief report. Virchows Arch. 424(2):225-227.

Alt, E.R., I. Sternlieb and S. Goldfischer. 1990. The cytopathology of metal overload. Int. Rev. Exp. Pathol. 31:165-188.

Aston, N., P. Morris, and S. Tanner. 1996. Retrorsine in breast milk influences copper handling in suckling rat pups. J. Hepatol. 25(5):748-755.

Aston, N.S., P.A. Morris, M.S. Tanner and S. Variend. 1998. An animal model for copper-associated cirrhosis in infancy. J. Pathol. 186(2):215-221.

ATSDR (Agency for Toxic Substances and Disease Registry). 1990. Toxicological Profile for Copper. PB91-180513. Prepared by Syracuse Research Corporation under Subcontract No. ATSDR-88-0608-02 for the Agency of Toxic Substances and Disease Registry. ATSDR, U.S. Public Health Service, December.

Aulerich, R.J., R.K. Ringer, M.R. Bleavins, A. Napolitano. 1982. Effects of supplemental dietary copper on growth, reproductive performance and kit survival of standard dark mink and the acute toxicity of copper to mink. J. Anim. Sci. 55(2):337-343.

Auza, N. 1983. Copper in ruminants. Review [in French]. Ann. Rech. Vet. 14(1):21-37.

Baker, A., S. Gormally, R. Saxena, D. Baldwin, B. Drumm, J. Bonham, B. Portmann, A.P. Mowat. 1995. Copper-associated liver disease in childhood. J. Hepatol. 23(5):538-543.

Barash, A., Z. Shoham, R. Borenstein, and L. Nebel. 1990. Development of human embryos in the presence of a copper intrauterine device. Gynecol. Obstet. Invest. 29(3):203-206.

Barbour, B.H., M. Bischel, and D.E. Abrams. 1971. Copper accumulation in patients undergoing chronic hemodialysis. The role of cuprophan. Nephron 8(5):455-462.

Barceloux, D.G. 1999. Copper. J. Toxicol. Clin. Toxicol. 37(2):217-230.

Becker, T.W., G. Krieger, and I. Witte. 1996. DNA single and double strand breaks induced by aliphatic and aromatic aldehydes in combination with copper (II). Free Radic. Res. 24(5):325-32.

Bent, S. and K. Bohm. 1995. Copper-induced liver cirrhosis in a 13-month old boy [in German]. Gesundheitswesen 57(10):667-669.

Berg, R. and S. Lundh. 1981. Copper contamination of drinking water as a cause of diarrhea in children. Halsovardskontakt 1:6-10.

Bhunya, S.P. and P.C. Pati. 1987. Genotoxicity of an inorganic pesticide, copper sulphate in mouse in vivo test system. Cytologia 52:801-808.

Bingham, M.J., T.J. Ong, K.H. Summer, R.B. Middleton and H.J. McArdle. 1998. Physiologic function of the Wilson disease gene product, ATP7B. Am. J. Clin. Nutr. 67(5 Suppl.):982S-987S.

Blomfield, J. 1969. Copper contamination in exchange transfusions. Lancet 1:731-732.

Blomfield, J., S.R. Dixon and D.A. McCredie. 1971. Potential hepatotoxicity of copper in recurrent hemodialysis. Arch. Intern. Med. 128(4):555-560.

Blomfield, J., J. McPherson, and C.R. George. 1969. Active uptake of copper and zinc during haemodialysis. Br. Med. J. 1(650):141-145.

Bohré, G.F., J. Huisman, H.F.L. Liefferink. 1965. Acute copper poisoning aboard a ship [in Dutch]. Ned. Tijdschr. Geneeskd 109:978-979.

Boyden, R., V.R. Potter and C.A. Elvehjem. 1938. Effect of feeding high levels of copper to albino rats. J. Nutr. 15:397-402.

Brewer, G.J. In press. Editorial: Is heterozygosity for a Wilson's disease gene defect an important underlying cause of infantile and childhood copper toxicosis syndromes? J. Trace Elem. Exp. Med.

Brewer, G.J. and V. Yuzbasiyan-Gurkan. 1992. Wilson disease. Medicine 71(3):139-164.

Brewer, G.J., W. Schall, R. Dick, V. Yuzbasiyan-Gurkan, M. Thomas and G. Padgett. 1992. Use of 64copper measurements to diagnose canine copper toxicosis. J. Vet. Intern. Med. 6(1):41-43.

Buchanan, S.D., R. Diseker, T. Sinks, J. Daniel, and T. Floodman. 1998. Evaluating Human Gastrointestinal Irritation From Copper in Drinking Water, Lincoln Nebraska, 1994. Draft. National Center for Environmental Health, Centers for Disease Control and Prevention, Atlanta, GA, and Division of Drinking Water and Environmental Sanitation, Nebraska Department of Health, Lincoln, NE.

Carlton, W.W., and P.S. Price. 1973. Dietary copper and the induction of neoplasms in the rat by acetylaminofluorene and dimethylnitrosamine. Food Cosmet. Toxicol. 11(5):827-840.

Cavallo, F., M. Gerber, E. Marubini, S. Richardson, A. Barbieri, A. Costa, A. DeCarli, H. Pujol. 1991. Zinc and copper in breast cancer. A joint study in northern Italy and southern France. Cancer 67(3):738-745.

CDC (Centers for Disease Control and Prevention). 1974. Acute Copper Poisoning—Arizona. MMWR (Nov. 23):407.

CDC (Centers for Disease Control and Prevention). 1975. Acute Copper Poisoning—Pennsylvania. MMWR (March 15):99.
CDC (Centers for Disease Control and Prevention). 1977. Outbreak of Acute Gastroenteritis Due to Copper Poisoning. MMWR (July 8):218, 223.
CDC (Centers for Disease Control and Prevention). 1996. Surveillance for Waterborne-Disease Outbreaks—United States, 1993-1994. MMWR 45(SS-1):12-13.
Cetinkaya, N., D. Cetinkaya and M. Yuce. 1988. Serum copper, zinc levels, and copper. Zinc ratio in healthy women and women with gynecological tumors. Biol. Trace Elem. Res. 18:29-38.
Chowdhury, A.K., S. Ghosh and D. Pal. 1961. Acute copper sulphate poisoning. J. Indian Med. Assoc. 36: 330-336.
Chugh, K.S., P.C. Singhal, and B.K. Sharma. 1975. Methemoglobinemia in acute copper sulfate poisoning. Ann. Intern. Med. 82(2):226-227.
Chugh, K.S., B.K. Sharma, P.C. Singhal, K.C. Das and B.N. Datta. 1977. Acute renal failure following copper sulphate intoxication. Postgrad. Med. J. 53(615):18-23.
Chuttani, H.K., P.S. Gupta, S. Gulati and D.N. Gupta. 1965. Acute copper sulfate poisoning. Am. J. Med. 39(5):849-854.
Clark, J.B. 1953. The mutagenic action of various chemicals on *Micrococcus aureus*. Proc. Okla. Acad. Sci. 34:114-118.
Coates, R.J., N.S. Weiss, J.R. Daling, R.L. Rettmer, and G.R. Warnick. 1989. Cancer risk in relation to serum copper levels. Cancer Res. 49(15):4353-4356.
Csermely, P., M. Szamel, K. Resch and J. Somogyi. 1988. Zinc can increase the activity of protein kinase C and contributes to its binding to plasma membranes in T lymphocytes. J. Biol. Chem. 263(14):6487-6490.
Cuthbert, J.A. 1995. Wilson's disease: a new gene and an animal model for an old disease. J. Invest. Med. 43(4):323-336.
Dabek, J.T., M. Hyvonen-Dabek, M. Harkonen and H. Adlercreutz. 1992. Evidence for increased non-ceruloplasmin copper in early-stage human breast cancer serum. Nutr. Cancer 17(2):195-201.
Dagenais, S.L., M. Guevara-Fujita, R. Loechel, A.C. Burgess, D.E. Miller, V. Yuzbaysian-Gurkan, G.J. Brewer, T.W. Glover. 1999. The canine copper toxicosis locus is not syntenic with ATP7B or ATX1 and maps to a region showing homology to human 2p21. Mamm. Genome 10(7):753-756.
Deng, D.X., S. Ono, J. Koropatnick and M.G. Cherian. 1998. Metallothionein and apoptosis in the toxic milk mutant mouse. Lab. Invest. 78(2):175-83.
Denizeau, F. and M. Marion. 1989. Genotoxic effects of heavy metals in rat hepatocytes. Cell Biol. Toxicol. 5(1):15-25.

de Vries, D.J., R.B. Sewell, and P.M. Beart. 1986. Effects of Copper on dopaminergic function in the rat corpus striatum. Exp. Neurol. 91(3):546-558.

Dick, A.T., D.W. Dewey, J.M. Gawthorne. 1975. Thiomolybdates. and the copper-molybdenum-sulfur interaction in ruminant nutrition. J. Agr. Sci. 85(3):567-568.

Donohue, J. 1997. New ideas after five years of the lead and copper rule: a fresh look at the MCLG for copper. Pp. 265-272 in Advances in Risk Assessment of Copper in the Environment, G.E. Lagos and R. Badilla-Ohlbaum, eds. Santiago, Chile: Catholic University of Chile.

Downey, J.S., C.D. Bingle, S. Cottrell, N. Ward, D. Churchman, M. Dobrota, and C.J. Powell. 1998. The LEC rat possesses reduced hepatic selenium, contributing to the severity of spontaneous hepatitis and sensitivity to carcinogenesis. Biochem. Biophys. Res. Commun. 244(2): 463-7.

DuBois, R.S., O. Rodgerson, G. Martinau, G. Shroter, G. Giles, J. Lilly, C.G. Halgrimson, and T.E. Starzl. 1971. Orthotopic liver transplantation for Wilson's disease. Lancet 13(March):505-508.

EPA (U.S. Environmental Protection Agency). 1987. Summary Review of the Health Associated with Copper. Health Issue Assessment. EPA/600/8-87/001. Environmental Criteria and Assessment Office, U.S. Environmental Protection Agency, Cincinnati, OH.

Epstein, O., R. Spisni, S. Parbhoo, B. Woods, and T. Dormandy. 1982. The effect of oral copper loading and portasystemic shunting on the distribution of copper in the liver, brain, kidney, and cornea of the rat. Am. J. Clin. Nutr. 35(3):551-555.

Fairbanks, V.F. 1967. Copper sulfate-induced hemolytic anemia. Inhibition of glucose-6-phosphate dehydrogenase and other possible etiologic mechanisms. Arch. Intern. Med. 120(4):428-32.

Ferm, V.H., and D.P. Hanlon. 1974. Toxicity of copper salts in hamster embryonic development. Biol. Reprod. 11(1):97-101.

Follesa, P., A. Mallei, S. Floris, M.C. Mostallino, E. Sanna and G. Biggio. 1999. Increased abundance of GABAA receptor subunit mRNAs in the brain of Long-Evans Cinnamon rats, an animal model of Wilson's disease. Brain Res. Mol. Brain Res. 63(2):268-75.

Freedman, J.H., M.R. Ciriolo and J. Peisach. 1989. The role of glutathione in copper metabolism and toxicity. J. Biol. Chem. 264(10):5598-605.

Fuentealba, I. and S. Haywood. 1988. Cellular mechanisms of toxicity and tolerance in the copper-loaded rat. I. Ultrastructural changes in the liver. Liver. 8(6):372-380.

Ghosh, S. and V.P. Aggarwal. 1962. Accidental poisoning in childhood, with particular reference to kerosene. J. Indian Med. Assoc. 39(Dec.):635-639.

Glass, G.A., and A.A. Stark. 1997. Promotion of glutathione-gamma-glutamyl transpeptidase-dependent lipid peroxidation by copper and ceruloplasmin: the requirement for iron and the effects of antioxidants and antioxidant enzymes. Environ Mol Mutagen 29(1):73-80.

Goldstein, S. and G. Czapski. 1986. The role and mechanism of metal ions and their complexes in enhancing damage in biological systems or in protecting these systems from the toxicity of O2-. J. Free Radic. Biol. Med. 2(1):3-11.

Gooneratne, S.R., J.M. Howell and J.M. Gawthorne. 1981. An investigation of the effects of intravenous administration of thiomolybdate on copper metabolism in chronic Cu-poisoned sheep. Br. J. Nutr. 46(3): 469-480.

Griffin, A.J.B. 1951. Unusual case of copper poisoning. J. R. San. Inst. 71(Jan.):1-8.

Halliwell, B. 1989. Free radicals, reactive oxygen species and human disease: a critical evaluation with special reference to atherosclerosis. Br. J. Exp. Pathol. 70(6):737-757.

Hanna, L.A., J.M. Peters, L.M. Wiley, M.S. Clegg, C.L. Keen. 1997. Comparative effects of essential and nonessential metals on preimplantation mouse embryo development in vitro. Toxicology 116:123-131.

Hansen, M.J. and H.G. Stefan. 1984. Side effects of 58 years of copper sulfate of the Fairmont Lakes, Minnesota. Water Resour. Bull. 20(6):889-900.

Harris, Z.L.. and J.D. Gitlin. 1996. Genetic and molecular basis for copper toxicity. Am. J. Clin. Nutr. 63(5):836S-841S.

Harrison, J.W.E., S.E. Levin, and B. Trabin. 1954. The safety and fate of potassium sodium copper chlorophyllin and other copper compounds. J. Am. Pharm Assoc. 43:722-737.

Haywood, S. 1980. The effect of excess dietary copper on the liver and kidney of the male rat. J. Comp. Pathol. 90(2):217-232.

Haywood, S. 1985. Copper toxicosis and tolerance in the rat. I. Changes in copper content of the liver and kidney. J. Pathol. 145(2):149-158.

Haywood, S. and B. Comerford. 1980. The effect of excess dietary copper on plasma enzyme activity and on the copper content of the blood of the male rat. J. Comp. Pathol. 90(2):233-238.

Haywood, S. and M. Loughran. 1985. Copper toxicosis and tolerance in the rat. II. Tolerance—a liver protective adaptation. Liver 5(5):267-275.

Haywood, S., M. Loughran and R.M. Batt. 1985a. Copper toxicosis and tolerance in the rat. III. Intracellular localization of copper in the liver and kidney. Exp. Mol. Pathol. 43(2):209-219.

Haywood, S., J.J. Trafford and M. Loughran. 1985b. Copper toxicosis and tolerance in the rat: IV. Renal tubular excretion of copper. Br. J. Exp. Pathol. 66(6):699-707.

Haywood, S., I.C. Fuentealba, J. Foster and G. Ross. 1996. Pathobiology of copper-induced injury in Bedlington terriers: ultrastructural and microanalytical studies. Anal. Cell. Pathol. 10(3):229-241.

Holmes, N.G., M.E. Herrtage, E.J. Ryder and M.M. Binns. 1998. DNA marker C04107 for copper toxicosis in a population of Bedlington terriers in the United Kingdom. Vet. Rec. 142(14):351-352.

Holtzman, N.A., and R.H. Haslam. 1968. Elevation of serum copper following copper sulfate as an emetic. Pediatrics 42(1):189-193.

Holtzman, N.A., D.A. Elliott, and R.H. Heller. 1966. Copper intoxication. Report of a case with observations on ceruloplasmin. N. Engl. J. Med. 275(7):347-352.

Hopper, S.H., and H.S. Adams. 1958. Copper poisoning from vending machines. Public Health Rep. 73(10):910-914.

Horslen, S.P., M.S. Tanner, T.D. Lyon, G.S. Fell, and M.F. Lowry. 1994. Copper associated childhood cirrhosis. Gut 35(10):1497-1500.

Howell, J.S. 1958. The effect of copper acetate on *p*-dimethylaminoazobenzene carcinogenesis in the rat. Br. J. Cancer 12(4):594-608.

Hultgren, B.D., J.B. Stevens and R.M. Hardy. 1986. Inherited, chronic, progressive hepatic degeneration in Bedlington terriers with increased liver copper concentrations: clinical and pathologic observations and comparison with other copper-associated liver diseases. Am. J. Vet. Res. 47(2):365-377.

Hurley, L.S. and C.L. Keen. 1979. Teratogenic effects of copper. Pp. 33-56 in Copper in the Environment. Part II: Health Effects, J.O. Nriagu, ed. New York: John Wiley & Sons.

Iyer, V.N. and W. Szybalski. 1958. Two simple methods for the detection of chemical mutagens. Appl. Microbiol. 6(1):23-29.

IPCS (International Programme on Chemical Safety). 1998. Copper. Environmental Health Criteria 200. Geneva: WHO.

Jantsch, W., K. Kulig and B.H. Rumack. 1985. Massive copper sulfate ingestion resulting in hepatotoxicity. Clin. Toxicol. 22(6):585-588.

Kanematsu, N., M. Hara and T. Kada. 1980. Rec assay and mutagenicity studies on metal compounds. Mutat. Res. 77(2):109-16.

Keen, C.L. 1996. Teratogenic effects of essential trace metals: deficiencies and excesses. Pp. 977-1001 in Toxicology of Metals, L.W. Chang, L. Magos, and T. Suzuki, eds. New York: CRC Press.

Keen, C.L., B. Lönnerdal and L.S. Hurley. 1982. Teratogenic effects of copper deficiency and excess. Pp.109-121 in Inflammatory Diseases and Copper. J..R..J. Sorenson, ed. Clifton, NJ: Humana Press.

Kitaura, K., Y. Chone, N. Satake, A. Akagi, T. Ohnishi, Y. Suzuki and K. Izumi. 1999. Role of copper accumulation in spontaneous renal carcinogenesis in Long-Evans Cinnamon rats. Jpn. J. Cancer Res. 90(4):385-392.

Klein W.J., E.N. Metz and A.R. Price. 1972. Acute copper intoxication. A

hazard of hemodialysis. Arch. Intern. Med. 129(4):578-582.
Klein, D., J. Lichtmannegger, U. Heinzmann, J. Müller-Hocker, S. Michaelsen, K.H. Summer. 1998. Association of copper to metallothionein in hepatic lysosomes of Long-Evans cinnamon (LEC) rats during the development of hepatitis. Eur. J. Clin. Invest. 28(4):302-310.
Kline, R.D., V.W. Hays, and G.L. Cromwell. 1971. Effects of copper, molybdenum and sulfate on performance, hematology and copper stores in pigs and lambs. J. Anim. Sci. 33(4):771-779.
Knobeloch, L., M. Ziarnik, J. Howard, B. Theis, D. Farmer, H. Anderson, and M. Proctor. 1994. Gastrointestinal upsets associated with ingestion of copper-contaminated water. Environ. Health Perspect. 102(11):958-961.
Kodama, H. 1996. Genetic disorders of copper metabolism. Pp. 371-386 in Toxicology of Metals, L.W. Chang, L. Magos, and T. Suzuki, eds. Boca Raton, FL: CRC Press.
Koizumi, M., J. Fujii, K. Suzuki, T. Inoue, T. Inoue, J.M. Gutteridge and N. Taniguchi. 1998. A marked increase in free copper levels in the plasma and liver of LEC rats: an animal model for Wilson disease and liver cancer. Free Radic. Res. 28(5):441-450.
Kok, F.J., C.M. Van Duijn, A. Hofman, G.B. Van der Voet, F.A. De Wolff, C.H. Paays, and H.A. Valkenburg. 1988. Serum copper and zinc and the risk of death from cancer and cardiovascular disease. Am. J. Epidemiol. 128(2):352-359.
Lefkowitch, J.H., C.L. Honig, M.E. King, and J.W.C. Hagstrom. 1982. Hepatic copper overload and features of Indian childhood cirrhosis in an American sibship. N. Engl. J. Med. 307(5):271-277.
Le Van, J.H., and E.L. Perry. 1961. Copper poisoning on shipboard. Public Health Rep. 76:334.
Lecyk, M. 1980. Toxicity of cupric sulfate in mice embryonic development. Zool. Pol. 28:101-105.
Li, Y., and M.A. Trush. 1993. DNA damage resulting from the oxidation of hydroquinone by copper: role for a Cu(II)/Cu(I) redox cycle and reactive oxygen generation. Carcinogenesis 14(7):1303-11.
Li, Y., Y. Togashi, S. Sato, T. Emoto, J.H. Kang, N. Takeichi, H. Kobayashi, Y. Kojima, Y. Une and J. Uchino. 1991. Spontaneous hepatic copper accumulation in Long-Evans Cinnamon rats with hereditary hepatitis. A model of Wilson's disease. J. Clin. Invest. 87(5):1858-1861.
Lim, C.T. and K.E. Choo. 1979. Wilson's disease in a 2-year-old child. J. Singapore Paediat. Soc. 11(1&2):99-102.
Lindquist, R.R. 1967. Studies on the pathogenesis of hepatolenticular degeneration. I. Acid phosphatase activity in copper-loaded rat livers. Am. J. Pathol. 51(4):471-481.
Lindquist, R.R. 1968. Studies on the pathogenesis of hepatolenticular

degeneration. 3. The effect of copper on rat liver lysosomes. Am. J. Pathol. 53(6):903-927.

Llewellyn, G.C., E.A. Floyd, G.D. Hoke, L.B. Weekley, and T.D. Kimbrough. 1985. Influence of dietary aflatoxin, zinc, and copper on bone size, organ weight, and body weight in hamsters and rats. Bull. Environ. Contam. Toxicol. 35(2):149-156.

Loeb, L.A., E.A. James, A.M. Waltersdorph and S.J. Klebanoff. 1988. Mutagenesis by the autoxidation of iron with isolated DNA. Proc. Natl. Acad. Sci. (USA) 85(11):3918-22.

Logue, J.N., M.D. Koontz, and M.A. Hattwick. 1982. A historical prospective mortality study of workers in copper and zinc refineries. J. Occup. Med. 24(5):398-408.

Low, B., J.M. Donahue and C.B. Bartley. 1996. A Study on Backflow Prevention Associated with Carbonators. Final Report. NSF International, Ann Arbor, MI.

Ludwig, J., G.H. Farr, D.K. Freese, and I. Sternlieb. 1996. Chronic hepatitis and hepatic failure in a 14-year-old girl. Hepatology 22(6):1874-1879.

Luo, S.Q., M.C. Plowman, S.M. Hopfer, F.W. Sunderman, Jr. 1993. Embryotoxicity and teratogenicity of Cu2+ and Zn2+ for Xenopus laevis, assayed by the FETAX procedure. Ann. Clin. Lab Sci. 23(2):111-20.

Lyle, W.H., J.E. Payton and M. Hui. 1976. Haemodialysis and copper fever. Lancet 1(7973):1324-1325.

Ma, Y., D. Zhang, T. Kawabata, T. Kiriu, S. Toyokuni, K. Uchida and S. Okada. 1997. Copper and iron-induced oxidative damage in non-tumor bearing LEC rats. Pathol. Int. 47(4):203-208.

Maeda, Y., T. Taira, K. Haraguchi, K. Hirose, A. Kazusaka and S. Fujita. 1997. Activation of serum response factor in the liver of Long-Evans Cinnamon (LEC) rat. Cancer Lett. 119(2):137-141.

Maggiore, G., C. De Giacomo, F. Sessa, and G.R. Burgio. 1987. Idiopathic hepatic copper toxicosis in a child. J. Pediatr. Gastroenterol. Nutr. 6(6):980-983.

Makale, M.T., and G.L. King. 1992. Surgical and pharmacological dissociation of cardiovascular and emetic responses to intragastric CuSO4. Am. J. Physiol. 263(2 Pt 2):R284-R291.

Malins, D.C., N.L. Polissar and S.J. Gunselman. 1996. Tumor progression to the metastatic state involves structural modifications in DNA markedly different from those associated with primary tumor formation. Proc. Natl. Acad. Sci. (USA) 93(24):14047-52.

Manzler, A.D. and A.W. Schreiner. 1970. Copper-induced acute hemolytic anemia. A new complication of hemodialysis. Arch. Intern. Med. 73(3):409-412.

Marois, M., and M. Bovet. 1972. Effect of copper ions on pregnancy in

rats and rabbits [in French]. CR Soc. Biol. 166(10):1237-1240.
Marzin, D.R. and H.V. Phi. 1985. Study of the mutagenicity of metal derivatives with *Salmonella typhimurium* TA102. Mutat. Res. 155(1-2): 49-51.
Mason, J. 1990. The biochemical pathogenesis of molybdenum-induced copper deficiency in ruminants: towards the final chapter. Irish Vet. J. 43(1):18-21.
Massie, H.R. and V.R. Aiello. 1984. Excessive intake of copper: influence on longevity and cadmium accumulation in mice. Mech. Ageing Dev. 26(2-3):195-203.
Matsui, S. 1980. Evaluation of a Bacillus subtilis rec-assay for detection of mutagens which may occur in water environments. Water Res. 14(11):1613-1619.
Matsumoto, A., R. Hanayama, M. Nakamura, K. Suzuki, J. Fujii, H. Tatsumi, and N. Taniguchi. 1998. A high expression of heme oxygenase-1 in the liver of LEC rats at the stage of hepatoma: the possible implication of induction in uninvolved tissue. Free Radic. Res. 28(4): 383-391.
Matter, B.J., J. Pederson, G. Psimenos and R.D. Lindeman. 1969. Lethal copper intoxication in hemodialysis. Trans. Am. Soc. Artif. Intern. Organs 15:309-315.
McMullen, W. 1971. Copper contamination of soft drinks from bottle pourers. Health Bull. (Edinb.) 29(2):94-96.
McNatt, E.N., W.R. Campbell Jr., and B.C. Callahan. 1971. Effects of dietary copper loading on livers of rats. I. Changes in subcellular acid phosphatases and detection of an additional acid p-nitrophenylphosphatase in the cellular supernatant during copper loading. Am. J. Pathol. 64(1):123-144.
Miller, D.M., G.R. Buettner and S.D. Aust. 1990. Transition metals as catalysts of "autoxidation" reactions. Free Radic. Biol. Med. 8(1):95-108.
Mittal, S.R. 1972. Oxyhaemoglobinuria following copper sulphate poisoning: a case report and a review of the literature. Forensic Sci. 1(2):245-248.
Montaser, A., C. Tetreault, and M. Linder. 1992. Comparison of copper binding components in dog serum with those in other species. Proc. Soc. Exp. Biol. Med. 200(3):321-329.
Moore, G.S., and E.J. Calabrese. 1980. G6PD-deficiency: A potential high-risk group to copper and chlorite ingestion. J. Environ. Pathol. Toxicol. 4(2-3):271-279.
Mori, M., A. Hattori, M. Sawaki, N. Tsuzuki, N. Sawada, M. Oyamada, N. Sugawara, and K. Enomoto. 1994. The LEC rat: a model for human hepatitis, liver cancer, and much more. Am. J. Pathol. 144(1):200-204.

Moriya, M., T. Ohta, K. Watanabe, T. Miyazawa, K. Kato, and Y. Shirasu. 1983. Further mutagenicity studies on pesticides in bacterial reversion assay systems. Mutat. Res. 116(3-4):185-216.

Mudassar, S., K.I. Andrabi, M. Khullar, N.K. Ganguly and B.N. Walia. 1992. Effect of exogenous copper on lipid peroxidation in rat hepatocytes. Possible involvement of protein kinase C. J. Pharm. Pharmacol. 44(7):609-611.

Müller, T., H. Feichtinger, H. Berger and W. Müller. 1996. Endemic tyrolean infantile cirrhosis: an exogenetic disorder. Lancet 347(9005):877-880.

Müller, T.H., W. Müller and H. Feichtinger. 1998. Idiopathic copper toxicosis. Am. J. Clin. Nutr. 67(5 Suppl.):1082S-1086S.

Müller-Höcker, J., U. Meyer, B. Wiebecke, G. Hübner, R. Eife, M. Kellner, P. Schramel. 1988. Copper storage disease of the liver and chronic dietary copper intoxication in two further German infants mimicking Indian childhood cirrhosis. Pathol. Res. Pract. 183(1):39-45.

Müller-Höcker, J., M. Weiss, U. Meyer, P. Schramel, B. Wiebecke, B.H. Belohradsky, G. Habner. 1987. Fatal copper storage disease of the liver in a German infant resembling Indian childhood cirrhosis. Virchows. Arch. A. Pathol. Anat. Histopathol. 411(4):379-385.

Muramatsu Y., T. Yamada, D.H. Moralejo, Y. Suzuki, and K. Matsumoto. 1998. Fetal copper uptake and a homolog (Atp7b) of the Wilson's disease gene in rats. Res. Commun. Mol. Pathol. Pharmacol. 101(3):225-231.

Murthy, R.C., S. Lal, D.K. Saxena, G.S. Shukla, M.M. Ali, and S.V. Chandra. 1981. Effect of manganese and copper interaction on behavior and biogenic amines in rats fed a 10% casein diet. Chem. Biol. Interact. 37(3):299-308.

Myers, B.M., F.G. Prendergast, R. Holman, S.M. Kuntz, and N.F. Larusso. 1993. Alterations in hepatocyte lysosomes in experimental hepatic copper overload in rats. Gastroenterology 105(6):1814-1823.

Nagano, K., K. Nakamura, K.I. Urakami, K. Umeyama, H. Uchiyama, K. Koiwai, S. Hattori, T. Yamamoto, I. Matsuda and F. Endo. 1998. Intracellular distribution of the Wilson's disease gene product (ATPase7B) after in vitro and in vivo exogenous expression in hepatocytes from the LEC rat, an animal model of Wilson's disease. Hepatology 27(3):799-807.

Nakamura, K., F. Endo, T. Ueno, H. Awata, A. Tanoue and I. Matsuda. 1995. Excess copper and ceruloplasmin biosynthesis in long-term cultured hepatocytes from Long-Evans Cinnamon (LEC) rats, a model of Wilson disease. J. Biol. Chem. 270(13):7656-7660.

Nicholas, P.O. 1968. Food poisoning due to copper in the morning tea. Lancet 2(7558):40-42.

Nieminen, A.L.. and J.J. Lemasters. 1996. Hepatic injury by metal accu-

mulation. Pp. 887-899 in Toxicology of Metals, L.W. Chang, ed. Boca Raton, FL.: CRC Press.
Nishioka, H. 1975. Mutagenic activities of metal compounds in bacteria. Mutat. Res. 31(3):185-9.
NRC (National Research Council). 1977. Copper. Washington, D.C.: National Academy of Sciences.
NTP (National Toxicology Program). 1993. NTP Technical Report on Toxicity studies of Cupric Sulfate (CAS No. 7758-99-8) Administered in Drinking Water and Feed to F344/N Rats and B6C3F1 Mice. NTIS PB94-120870.
O'Donohue, J., M.A. Reid, A. Varghese, B. Portmann, R. Williams. 1993. Micronodular cirrhosis and acute liver failure due to chronic copper self-intoxication. Eur. J. Gastroenterol. Hepatol. 5:561-562.
Ogra, Y. and K.T. Suzuki. 1998. Targeting of tetrathiomolybdate on the copper accumulating in the liver of LEC rats. J. Inorg. Biochem. 70(1): 49-55.
Olivier, P and D. Marzin. 1987. Study of the genotoxic potential of 48 inorganic derivatives with the SOS chromotest. Mutat. Res. 189(3):263-9.
Overvad, K., D.Y. Wang, J. Olsen, D.S. Allen, E.B. Thorling, R.D. Bulbrook, and J.L. Hayward. 1993. Copper in human mammary carcinogenesis: a case-cohort study. Am. J. Epidemiol. 137(4):409-414.
Owen, C.A. Jr., E.J. Bowie, J.T. McCall, and P.E. Zollman. 1980. Hemostasis in the copper-laden Bedlington terrier: a possible model of Wilson's disease. Haemostasis 9(3):160-166.
Percival T. 1784. A history of the fatal effects of pickles impregnated with copper, together with observations on that mineral poison. Med. Trans. R. Coll. Phys. (London) 3:80-95.
Pizarro, F., M. Olivares, R. Uauy, P. Contreras, A. Rebelo, and V. Gidi. 1999. Acute gastrointestinal effects of graded levels of copper in drinking water. Environ. Health Perspect. 107(2):117-121.
Prasad, R., G. Kaur, R. Mond, B.N. Walia. 1998. Identification of a novel copper-binding protein from the liver of Indian childhood cirrhosis: purification and physicochemical characterization. Pediatr. Res. 44(5):673-681.
Prasad, M.P., T.P. Krishna, S. Pasricha, K. Krishnaswamy, and M.A. Quereshi. 1992. Esophageal cancer and diet—a case-control study. Nutr. Cancer 18(1):85-93.
Rana, S.V.S. and A. Kumar. 1980. Biological haematological and histological observations in copper poisoned rats. Ind. Health. 18:9-17.
Rauch, H. 1983. Toxic milk, a new mutation affecting cooper metabolism in the mouse. J. Hered. 74(3):141-4.
Roberts, R.H. 1956. Hemolytic anemia associated with copper sulfate poisoning. Mississipi Doctor 33(March):292-294.
Roberts, L.F., B. Ashby, E.G. Hallock and M.McGeehin. 1996. Ability of

Household Consumers to Tolerate Elevated Copper Levels in Drinking Water: Delaware, 1996. Draft. National Center for Environmental Health, Centers for Disease Control and Prevention, Atlanta, GA.

Ross, A.I. 1955. Vomiting and diarrhea due to copper in stewed apples. Lancet (July 9):87-88.

Rui, M. and K.T. Suzuki. 1997. Copper in plasma reflects its status and subsequent toxicity in the liver of LEC rats. Res. Commun. Mol. Pathol. Pharmacol. 98(3):335-346.

Runner, M. N., and J. R. Miller. 1956. Congenital deformity in the mouse as a consequence of fasting. Anat. Rec. 124:437-438.

Saito, T., T. Nagao, M. Okabe, K. Saito. 1996. Neurochemical and histochemical evidence for an abnormal catecholamine metabolism in the cerebral cortex of the Long-Evans Cinnamon rat before excessive copper accumulation in the brain. Neurosci. Lett. 216(3):195-198.

Saito, R., Y. Suehiro, H. Ariumi, K. Migita, N. Hori, T. Hashiguchi, M. Sakai, M. Saeki, Y. Takano, and H. Kamiya. 1998. Anti-emetic effects of a novel NK-1 receptor antagonist HSP-117 in ferrets. Neurosci. Lett. 254(3):169-172.

Salmon, M.P., and T. Wright. 1971. Chronic copper poisoning presenting as pink disease. Arch. Dis. Child. 46(245):108-110.

Sanghvi, L.M., R. Sharma, S.N. Misra and K.C. Samuel. 1957. Sulfhemoglobinemia and acute renal failure after copper sulfate poisoning. Report of two fatal cases. Arch. Pathol. 63:172-175.

Scheinberg, I.H. and I. Sternlieb. 1984. Wilson's disease. Major Problems in Internal Medicine. Vol. 23. Philadelphia: W.B. Saunders.

Scheinberg, I.H., and I. Sternlieb. 1996. Wilson disease and idiopathic copper toxicosis. Am. J. Clin. Nutr. 63(5):842S-845S.

Semple, A.B., W.H. Parry and D.E. Phillips. 1960. Acute copper poisoning. An outbreak traced to contaminated water from a corroded geyser. Lancet 2:700-701.

Shanaman, J.E. 1972. Report of One Year Chronic Oral Toxicity of Copper Gluconate W10219A in Beagle Dogs. Research Rep. No. 955-0353. Morris Plains, NJ: Warner Lambert Research Institute.

Shanaman, J.E., F.X. Wazeter, E.I. Goldenthal. 1972. One-Year Chronic Oral Toxicity of Copper Gluconate W/02/09A in Beagle Dogs. Research Rep. No. 955-0353. Morris Plains, NJ: Warner Lambert Research Institute.

Singh, M.M. and G. Singh. 1968. Biochemical changes in blood in cases of acute copper sulphate poisoning. J. Indian Med. Assoc. 50(12):449-554.

Sokol, R.J., M.W. Devereaux, M.G. Traber and R.H. Shikes. 1989. Copper toxicity and lipid peroxidation in isolated rat hepatocytes: effect of vitamin E. Pediatr. Res. 25(1):55-62.

Spitalny, K.C., J. Brondum, R.L. Vogt, H.E. Sargent, and S. Kappel. 1984.

Drinking water induced copper intoxication in a Vermont family. Pediatrics 74(6):1103-1106.
Stein, R.S., D. Jenkins, and M.E. Korns. 1976. Death after use of cupric sulfate as an emetic [letter]. JAMA 235(8):801.
Steinebach, O.M. and H.T. Wolterbeek. 1994. Role of cytosolic copper, metallothionein and glutathione in copper toxicity in rat hepatoma tissue culture cells. Toxicology 92(1-3):75-90.
Stenhammer, L. 1979. Copper intoxication: a differential diagnosis of diarrhea in children [in Swedish]. Lakartidningen 76(30-31):2618-2620.
Stoner, G.D., M.B. Shimkin, M.C. Troxell, T.L. Thompson, and L.S. Terry. 1976. Test for carcinogenicity of metallic compounds by the pulmonary tumor response in strain A mice. Cancer Res. 36(5):1744-1747.
Suttle, N.F. and C.F. Mills. 1966. Studies of the toxicity of copper to pigs. I. Effects of oral supplements of zinc and iron salts on the development of copper toxicosis. Br. J. Nutr. 20(2):135-148.
Suzuki, K.T. 1995. Disordered copper metabolism in LEC rats, an animal model of Wilson disease: roles of metallothionein. Res. Commun. Mol. Pathol. Pharmacol. 89(2):221-240.
Suzuki, K.T., M. Rui, J. Ueda and T. Ozawa. 1996. Production of hydroxyl radicals by copper-containing metallothionein: roles as prooxidant. Toxicol. Appl. Pharmacol. 141(1):231-237.
Tachibana, K. 1952. Pathological transition and functional vicissitude of liver during formation of cirrhosis by copper. Nagoya J. Med Sci. 15:108-114.
Tanner, M.S. 1998. Role of copper in Indian childhood cirrhosis. Am. J. Clin. Nutr. 67(5 Suppl.):1074S-1081S.
Tanner, M.S., and A.R. Mattocks. 1987. Hypothesis: plant and fungal biocides, copper and Indian childhood liver disease. Ann. Trop. Pediatr. 7(4):264-269.
Tanner, M.S., A.H. Kantarjian, S.A. Bhave and A.N. Pandit. 1983. Early introduction of copper-contaminated animal milk feeds as a possible cause of Indian childhood cirrhosis. Lancet 2(8357):992-995.
Terada, K., N. Aiba, X.L. Yang, M. Iida, M. Nakai, N. Miura and T. Sugiyama. 1999. Biliary excretion of copper in LEC rat after introduction of copper transporting P-type ATPase, ATP7B. FEBS Lett. 448(1):53-56.
Terada, K., M.L. Schilsky, N. Miura and T. Sugiyama. 1998. ATP7B (WND) protein. Int. J. Biochem. Cell Biol. 30(10):1063-1067.
Tkeshelashvili, L,K,, T. McBride, K. Spence and L.A. Loeb. 1991. Mutation spectrum of copper-induced DNA damage. J. Biol. Chem. 266(10):6401-6406.
van de Sluis, B.J., M. Breen, M. Nanji, M. van Wolferen, P. de Jong, M.M. Binns, P.L. Pearson, J. Kuipers, J. Rothuizen, D.W. Cox, C. Wijmenga,

and B.A. van Oost. 1999. Genetic mapping of the copper toxicosis locus in Bedlington terriers to dog chromosome 10, in a region syntenic to human chromosome region 2p13-p16. Hum. Mol. Genet. 8(3):501-507.

von Rosen, G. 1964. Mutations induced by action of metal ions in Pisum II. Hereditas 51(1):89-134.

Vulpe, C.D. and S. Packman. 1995. Cellular copper transport. Annu. Rev. Nutr. 15:293-322.

Wahal, P.K., V.P. Mittal, and O.P. Bansal. 1965. Renal complications in acute copper sulphate poisoning. Indian Pract. 18:807-812.

Walker-Smith, J. and J. Blomfield. 1973. Wilson's disease or chronic copper poisoning? Arch. Dis. Child. 48(6):476-479.

Walsh, F.M., F.J. Crosson, M. Bayley, J. McReynolds, and B.J. Pearson. 1977. Acute copper intoxication. Pathophysiology and therapy, with a case report. Am. J. Dis. Child. 131(2):149-151.

Wang, S.C. and H.I. Borison. 1951. Copper sulphate emesis: study of afferent pathways from the gastrointestinal tract. Am. J. Physiol. 164:520-526.

Wapnir, R.A. 1998. Copper absorption and bioavailability. Am. J. Clin. Nutr. 67(5 Suppl.):1054S-1060S.

Wong, P.K. 1988. Mutagenicity of heavy metals. Bull. Environ. Contam. Toxicol. 40(4):597-603.

Wu, J., J.R. Forbes, H.S. Chen, and D.W. Cox. 1994. The LEC rat has a deletion in the copper transporting ATPase gene homologous to the Wilson disease gene. Nat. Genet. 7(4):541-545.

Wyllie, J. 1957. Copper poisoning at a cocktail party. Am. J. Publ. Health 47:617.

Yuzbasiyan-Gurkan, V., B.S. Halloran, Y. Cao, P. Ferguson, J Li, P.J. Venta, G.J. Brewer. 1997. Linkage of a microsatellite marker to the canine copper toxicosis locus in Bedlington terriers. Am. J. Vet. Res. 58(1):23-27.

6

Risk Characterization

INTAKE of low concentrations of copper in the diet can result in deficiency, and high concentrations can result in toxicity. In the general population, there is a range of acceptable intakes that will meet copper requirements and pose no risk of toxicity (Figure 6-1). Generally, copper intake through diet appears to fall within this range for the average, normal, healthy individual. Typically, only a small fraction of an individual's intake of copper derives from drinking water; thus, drinking water should not be relied upon as an important source to meet daily copper requirements. On the other hand, drinking corrosive waters held in copper plumbing can result in copper excess, and the potential for copper toxicity is a concern in that case.

This chapter considers copper concentrations in drinking water that might produce copper excess and provides guidance on the establishment of the maximum contaminant level goal (MCLG). For those individuals with abnormal copper homeostasis, at vulnerable ages, or with already high copper concentrations from occupational exposures, the range of acceptable copper intake through water and diet can be narrow. This chapter also evaluates the prevalence of sensitive populations and the degree to which copper in drinking water might contribute to copper excess in individuals in those populations.

COPPER DEFICIENCY

Severe copper deficiency, characterized by bone abnormalities, severe anemia, compromised immune function, loss of skin, and growth retarda-

128   COPPER IN DRINKING WATER

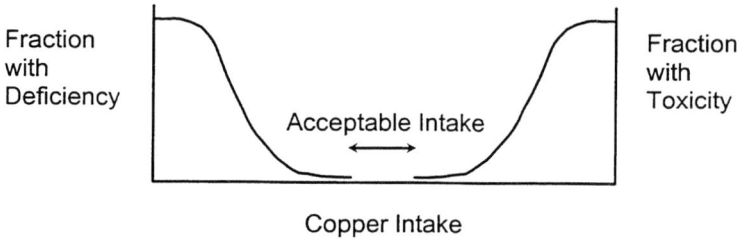

FIGURE 6-1   Range of acceptable copper intakes for the general population.

tion, is rarely observed clinically in the United States. Nonetheless, although nationwide surveys suggest that average intake concentrations within the population are at recommended concentrations, a substantial fraction of the population has intakes below currently recommended concentrations (Table 6-1). The significance of those low intakes remains to be determined.

Some populations are a particular concern:

• Pre-term infants have a lower prenatal accumulation of copper stores and thus can be at increased risk for developing copper deficiency during early infancy.
• Copper deficiency can be induced by select mineral supplements, particularly zinc.

Individuals taking zinc supplements in excess of the Recommended Daily Allowance (RDA) on a chronic basis might be at particular risk.

• Large numbers of the elderly appear to have dietary copper intakes below the recommended copper intake.
• A number of disease conditions, including diabetes and hypertension, are associated with low extrahepatic-tissue copper concentrations.
• Genetic disorders, such as occipital horn syndrome, might confer an increased risk for copper deficiency.

Several groups appear to be at risk for deficiency, but a substantial fraction of the population have intakes at or above the recommended level (Table 6-1). Therefore, the committee does not recommend redressing copper deficiency via the water supply and notes that the MCLG should not be established on the basis of copper deficiency.

TABLE 6-1 Copper Intake (mg/day)[a] from Food and Supplements versus the Estimated Safe and Adequate Daily Dietary Intake (ESADDI)

Age	Sex	Percentile								ESADDI[b] or recommended levels (mg/d)
		5	10	25	50	75	90	95	99	
2-6 mon	M&F	0.3	0.4	0.5	0.7	0.9	1.1	1.2	1.6	0.4 to 0.6 recommended
7-11 mon	M&F	0.3	0.4	0.5	0.7	0.9	1.2	1.3	1.7	0.6 to 0.7 recommended
1-3 yr	M&F	0.3	0.4	0.5	0.7	1.0	1.3	1.7	2.9	Recommended for 7-10-year-old children: 1 to 2
4-8 yr	M&F	0.59	0.67	0.80	0.95	1.14	1.36	1.61	3.06	
9-13 yr	F	0.64	0.72	0.86	1.04	1.26	1.54	1.84	3.23	
	M	0.88	0.94	1.05	1.21	1.41	1.61	1.78	3.13	
14-18 yr	F	0.64	0.75	0.89	1.08	1.32	1.64	1.96	3.32	
	M	0.79	0.89	1.11	1.42	1.80	2.28	2.71	3.56	
19-30 yr	F	0.77	0.83	0.95	1.12	1.38	1.82	3.03	3.84	ESADDI for adults: 1.5 to 3[c]
	M	1.37	1.43	1.56	1.69	1.86	2.12	3.55	4.44	
31-50 yr	F	0.72	0.81	0.95	1.17	1.52	2.32	3.09	4.19	
	M	0.89	1.03	1.29	1.61	2.09	2.93	3.67	4.87	
51-70 yr	F	0.61	0.68	0.84	1.07	1.48	2.92	3.25	4.22	
	M	0.75	0.87	1.09	1.43	1.98	3.00	3.65	5.02	
71+ yr	F	0.58	0.65	0.80	1.02	1.37	2.94	3.21	3.79	
	M	0.72	0.83	0.99	1.26	1.66	2.89	3.41	4.61	
Pregnant	F	0.71	0.82	1.07	1.62	3.11	4.03	4.39	5.56	

[a]Intake figures developed by Environ International Corporation and Iowa State University Department of Statistics from the NHANES III nationwide survey. Breast-feeding infants and children, and eight individuals reporting greater than 150 mg/day of copper from supplements excluded from the analysis.

[b]ESADDI values and recommended concentrations from the Food and Nutrition Board (FNB) (NRC 1989). The Institute of Medicine's (IOM) FNB is reviewing these values, and the reader should consult current IOM references for updated values.

[c]For comparison, the acceptable range of intake values from IPCS (1998) is given as 1 to "several but not many mg per day." "Several" is defined as more than 2 to 3 mg/day.

COPPER TOXICITY FROM SINGLE OR SHORT-TERM EXPOSURE

A wide range of copper toxicities have been observed subsequent to accidental or iatrogenic poisonings or suicide attempts: hepatic and renal failure, cirrhosis, hemolysis, vomiting, melena, hypotension, cardiovascular collapse, stupor, and coma (Chapter 5). Less severe acute copper toxicity, which can occur from corrosive water standing in copper plumbing and beverages or acidic foods stored or prepared in copper containers, is associated with nausea, vomiting, and diarrhea. Acute exposures at which gastrointestinal (GI) effects first appear are lower than those producing hepatotoxicity and other more severe acute effects.

Copper ions are generally more bioavailable in water than in food (Chapter 2). Because acute irritation of the GI tract is caused by the ionic form of copper, it is reasonable to suggest that GI irritation is more likely to be produced by drinking water than food. Consistent with that suggestion, the majority of reports on copper-induced GI irritation concern the ingestion of fluids high in this element.

Concentrations and doses associated with acute GI effects are detailed in Chapter 5. Human data are consistent with GI symptoms, such as nausea, abdominal pain, and vomiting, arising in sensitive individuals from drinking water and other fluids with copper concentrations greater than or equal to 3 mg/L (see Pizarro et al. 1999). Nausea typically occurs at lower concentrations than vomiting. In that study, some individuals were affected at 3 mg/L, but the majority reported no symptoms at 5 mg/L, indicating a range of sensitivity in the population. Nausea was noted as the sensitive effect in a larger experimental study (ICA, unpublished material, Oct. 13, 1999), and a range of sensitivity was also seen.

The MCLG is to be set at a concentration at which no known or expected adverse health effects occur and for which there is an adequate margin of safety. For noncancer end points, the approach typically applied to achieve that concentration is to determine a no-observed-adverse-effect level (NOAEL), by reviewing the health-effects literature and then dividing that level by dose adjustment and uncertainty factors (NRC 1977, 1994; IPCS 1994; EPA 1998). The adjustment factors applied account for human heterogeneity, study quality, study duration, and severity of effect. When based on the lowest-observed-adverse-effect level (LOAEL), an additional factor is applied to adjust to a level at which effects would not be expected to be observed. However, it is recognized that application of the standard regulatory approach for selection of safety factors for essential nutrients can result in levels below essential requirements.

The EPA (1991) established an MCLG for copper of 1.3 mg/L. The 1991 MCLG, explained in EPA (1994), is based on an analysis provided by EPA

(1987). The MCLG is based on a report (Wyllie 1957) of an episode of acute GI symptoms resulting from mixing alcoholic drinks in a copper-contaminated cocktail shaker. From a dose reconstruction, the author estimated that the dose associated with the lowest amount of cocktail consumed (half a glass, or 0.75 fluid ounces) resulting in symptoms was 5.3 mg. The level of 1.3 mg/L was recommended by EPA (1987) because it satisfied the nutritional requirements (noted by EPA (1987) to be 2-3 mg per day for adults and 1.5 to 2.5 mg per day for children) and because consumption of 2 L per day would result in intakes below the LOAEL (by a factor of 2). The EPA derivation thus uses total amount of daily copper as the dose metric for acute effects and a safety factor of 2.

Studies that have become available over the past decade have improved the basis for the establishment of the MCLG for copper. The recently published human experimental study of GI effects by Pizarro et al. (1999) indicates symptoms arising from copper exposure at approximately 3 mg/L of drinking water. The findings of the study are consistent with those from case reports, and because the study was controlled, it is preferred over them. With respect to the appropriate dose metric (discussed in Chapter 5) in studies of copper induction of the emetic response (e.g., Saito et al. 1998; Makale and King 1992), the response appears to be receptor mediated and a function of the effects of copper ion on the lining of the stomach. In human and experimental animals, the response occurs within a very short period after liquid is consumed, typically within a few minutes (Wang and Borison 1951; Pizarro et al. 1999). Thus in assessing the acute effects from exposure to copper in tap water, the water concentration appears to be of greater relevance than the amount of copper consumed over the day. Although additional work is needed to establish the relationship between concentration, volume consumed, and emetic response, the committee considers drinking-water concentration to be an appropriate dose metric for evaluating the MCLG with respect to the acute GI effects of copper.

In selecting adjustment uncertainty factors for setting the MCLG, issues to consider are the observations that (1) the effect is not severe and life threatening, and (2) the data supporting 3 mg/L as the concentration above which effects begin to appear are from controlled studies in humans and case reports, although much of the data are imprecise and limited. Limitations include the lack of measurement of confounders and biases and the small number of subjects in the published experimental study. Another consideration is the impact of taste on GI symptoms. The taste threshold for copper varies among individuals and depends on an individual's acuity. In controlled studies, Cohen et al. (1960), found that 50% of 15 to 20 taste panelists detected copper at concentrations of 6.6 mg/L and above in distilled water and 12.7 mg/L and above in spring water. They estimated the most sensitive 5% could taste copper at concentrations greater than 2.6 mg/L in distilled water and 5 mg/L in spring water. Thus, levels associated with the GI response and taste overlap, and taste is a potential confounder in copper studies.

Two additional issues need to be considered in establishing the overall adjustment and uncertainty factors for copper. First, the LOAELs from the controlled human studies appear to be at the lower part of the dose-response curve, where the majority of the population is nonresponsive. Second, copper is an essential micronutrient. Although the committee recommends against basing the MCLG on nutritional requirements, the fact that the compound is essential should limit the size of the overall factor selected.

Ultimately, the selection of an adjustment factor for the acute GI symptoms observed-effect level of 3 mg/L is a scientific and a policy issue. On the basis of the data on GI effects from acute and subacute exposures, IPCS (1998) established as the upper limit of the acceptable range of intake for total copper a value of "several but not many mg per day." "Several" is defined as more than 2 to 3 mg per day. WHO plans to re-evaluate copper in drinking water based on the IPCS (1998) analysis (Galal-Gorchev and Herrman 1996).

COPPER TOXICITY FROM CHRONIC EXPOSURE

The primary systemic effect of concern associated with chronic excess copper is liver toxicity (Chapter 5). In animal models, liver toxicity has been demonstrated in several species and there are case reports of liver disease in humans ingesting excessive concentrations of copper over a long period of time. In addition to liver pathology, Wilson disease suggests several other systemic end points for chronic copper toxicity. In theory, extrahepatic toxicity is subsequent to the accumulation of copper in the liver and results after cytosolic binding sites in the liver become saturated (Zucker and Gollan 1996). After that point, copper appears to be redistributed from the cytosol to lysosomes, and it can be released into systemic circulation. Copper can then accumulate in select extrahepatic sites, such as the kidney and brain. The neurological effects seen in Wilson-disease patients have not been observed in rodent studies of copper toxicity (see Chapter 5). Nonetheless, from Wilson disease and cases reports of accidental and intentional copper poisonings it can be concluded that the liver is a sensitive end point. An additional sensitive end point to consider that is unrelated to hepatic toxicity is GI irritation.

Sensitive Populations

Severe disorders of copper homeostasis, which occur with Wilson disease, result in hepatic toxicity and other copper toxicity, primarily involving the central nervous system. The prevalence of Wilson disease is small, reported to be roughly 1:40,000 live births (Chapter 6). However, because that estimate is based on autopsy data, it might be an underestimate. The actual prevalence of Wilson disease is likely to be considerably greater, by perhaps a factor of 4 or more. Those diagnosed with this serious illness are generally under a clinician's care. The Wilson-disease patient has limited ability to excrete copper, and copper intake through food and water is curtailed. Certain foods, such as liver and shellfish, that are high in copper should be avoided, and when the copper content in drinking-water sources is high (e.g., more than 0.1 mg/L), alternative sources are typically recommended. Thus, copper concentrations in drinking water for Wilson-disease patients are not managed through the MCLG process. However, in other groups, such as infants with altered copper metabolism, toxicity can occur when individuals in these groups consume drinking water with relatively high copper concentrations. These groups should be considered in establishing the MCLG.

Carriers of the Wilson-Disease Gene and Other Genetically Sensitive Groups

It is likely that a copper sensitivity gene contributes to the hepatic copper toxicity observed in infants and young children ingesting increased amounts of copper in milk and water. The current evidence is that manifestations of Tyrolean infantile cirrhosis (TIC), Indian childhood cirrhosis (ICC) and idiopathic copper toxicosis (ICT) involve both heredity and high copper intake (Muller et al. 1996 and 1998; Tanner 1998). It is a reasonable hypothesis that chronic ingestion of moderately increased amounts of copper produces disease in copper-susceptible genotypes.

Heterozygous carriers of the Wilson-disease gene might represent a susceptible group for copper hepatotoxicity. As evidence, under current environmental conditions in the United States, the heterozygous carriers accumulate copper and have abnormally high concentrations in the liver and urine (see Chapter 4). They can be defective in copper handling in the liver as evidenced by ^{64}Cu incorporation into ceruloplasmin (Brewer and Yuzbasiyan-Gurkan 1992). In addition, unidentified copper sensitivity genes might be responsible for the observed childhood copper toxicity syndromes (Muller et al., 1998; Tanner, 1998).

A heterozygote carrier rate of slightly greater than 1% corresponds to a prevalence rate of 1:40,000 for those homozygous for the Wilson-disease gene. The actual value might be considerably higher (on the order of 2%) if, as expected, the actual prevalence of Wilson disease is underestimated by a factor of approximately 4. Although Wilson heterozygote carriers likely differ in sensitivity, other genetic mutations might also increase copper retention. Thus, at least 1% of the population might be susceptible for increased copper retention on the basis of genetic susceptibility. Provided that increased copper retention confers increased risk of liver toxicity, the committee concludes that groups of this size should be taken into account in establishing the MCLG for chronic exposures.

Infants

Infants might represent a susceptible group for two reasons. First, newborns have approximately 3 times as much hepatic copper as adults. Those concentrations steadily decrease in early life and achieve adult concentrations after about 6 months (Keen 1996). Infants respond to higher copper intake by increasing fecal losses and decreasing percentage absorption, but they are not as efficient as adults at doing so, and cases of excess copper associated with genetic predisposition (e.g., TIC, and ICT) are most frequently observed in young children. Second, on a body-weight basis, infants drink considerably more water than adults, particularly those that are fed formula. The NHANES III (Table 6-1) and Continuing Survey of Food Intakes of Individuals (CSFII) indicate that intake of copper for the majority of infants is at adequate intake concentrations and above. Infants at higher percentile dietary intakes might be at a heightened risk for copper toxicity if intake of copper in drinking water is excessive. High exposures can especially result from use of first draw-water in preparing infant formula. Evidence from review of ICC and TIC cases, however, indicates a genetic predisposition to liver toxicity from elevated copper ingestion in infants and indicates that all infants are not equally sensitive (Muller et al. 1996; Tanner 1998).

Chronic Liver Disease

Copper accumulation in the liver is also associated with chronic cholestasis. Individuals with primary biliary cirrhosis, intrahepatic cholestasis in childhood, and extrahepatic biliary obstruction might be at increased risk. The extent to which the increased copper affects the already compromised liver is unknown.

Glucose-6-Phosphate Dehydrogenase Deficiency

Individuals with glucose-6-phosphate dehydrogenase deficiency (G6PD) have been hypothesized to be at increased risk for excess copper, because, in vitro, G6PD-deficient red blood cells are more susceptible than normal to hemolysis and damage from copper. However, as discussed in Chapter 5, copper bound to ceruloplasmin is not available for red-blood-cell toxicity. The relatively small amount of low-molecular-weight bound copper available is not likely to alter the survival of G6PD-deficient red blood cells. Thus, there is insufficient evidence that this group represents a susceptible group for chronic toxicity to copper from drinking water.

Overall, the fraction of the population with inherent sensitivity to copper due to genetic make-up, age, or disease state might be 1% or greater.

Implications for the MCLG

Copper is an essential nutrient subject to tight homeostatic control. Applying the standard regulatory approach for selecting safety factors and animal effect levels for essential elements and vitamins can result in guidance inconsistent with known nutritional requirements. Copper is one such case. Concentrations associated with effects observed in long-term animal studies are given in Table 6-2, along with adjustment and uncertainty factors traditionally applied. It is noteworthy that for the rat and rabbit, the LOAEL or NOAEL divided by the traditional adjustment and uncertainty factors results in concentrations below the dietary intake concentrations identified as safe and adequate for adults (1.5 to 3 mg/day, corresponding to roughly 20 to 40 µg/kg per day) and below the intake concentration of 75 µg/kg per day recommended for infants (NRC 1989). They also fall below the value identified by WHO (1996) as a probable average adult-based requirement (11 µg/kg-day) and safe minimum mean copper intake of populations (20 µg/kg-day) and below the values identified by WHO as normative requirements for infants (0-3 months, 50-80 µg/kg per day; 3-6 months, 40-70 µg/kg per day; 6-12 months, 40 µg/kg per day). The detailed pharmacokinetic and mechanistic information needed to perform a more refined safety assessment based on data from chronic exposure animal studies is not available. The chronic exposure animal studies provide qualitative support for findings in humans sensitive to copper. However, they provide little guidance for the establishment of an MCLG for copper.

Human case reports and series suggest a range of copper intakes associated with liver toxicity in sensitive individuals. Table 6-3 lists reports in the literature of increased copper exposure in cases of ICT, ICC, and TIC

TABLE 6-2 Copper-Effect Doses Observed in Long-Term Animal Studies

Species	Effect	LOAEL[a] (mg Cu/kg-d)	Traditional adjustment factor[b]	Reference
Mouse	Reduction in life span	42.5	1,000	Massie and Aiello 1984
Rat (90-d)	Hepatic; increase in SGOT activity	7.9	1,000-10,000	Epstein et al. 1982
Rat (92-d)	Renal Effects	16	10,000	NTP 1993
Rabbits	Marked hepatic toxicity	10	1,000	Tachibana 1952

[a]Effects were either observed at all doses or the only dose studied. Thus, NOAELs were not observed in the studies cited in the table.

[b]Typically, a factor of 10 is used to extrapolate from a dose associated with an effect (i.e., the LOAEL) to predict a NOAEL; a factor of 10 is used to extrapolate from animals to humans; a factor of 10 is used to account for variability within the human population; and a factor of 10 is used to extrapolate studies of insufficient duration to chronic duration.

Abbreviations: LOAEL, lowest-observed-adverse-effect level; NOAEL, no-observed-adverse-effect level.

and a case of self-exposure in an individual otherwise thought to be healthy until the time that signs of liver toxicity began to appear. Those reports are subject to imprecision in exposure ascertainment, but overall suggest that when formulas are made from drinking water containing 3 mg/L and above, genetically sensitive infants might be at increased risk of liver toxicity. It is difficult to ascertain copper concentrations in water and intake associated with toxicity because of varying copper concentrations due in large part to the flushing of the household system as water is used throughout the day. Although these estimates are far from exact, they nonetheless provide an indication of intakes that might cause hepato-toxicity to sensitive individuals upon chronic exposure.

The one case of chronic poisoning through self-exposure to copper supplements (O'Donohue et al., 1993) corresponds to a dose of 0.4-0.9 mg/kg per day, which is similar to crude estimates of dose for cases of ICC and TIC obtained by experimentally simulating copper storage and heating of formula milk (Table 6-3).

Division of the doses reported to induce hepatic toxicity in humans by traditional uncertainty and adjustment factors would result in a suggested maximal concentration of copper below those considered to be essential.

TABLE 6-3 Case Reports and Series of Toxicity Following Chronic Exposure to Copper

Effect and Subject	Exposure Circumstance and Copper Measurement	Reference
Young adult male; micronodular cirrhosis and liver failure; Wilson homozygosity and heterozygosity ruled out	Mineral supplementation of 30 mg/d for 2 yr, followed by 60 mg/d for 1 yr; estimated dose is 0.4 to 0.9 mg/kg per d	O'Donohue et al. 1993
Idiopathic copper toxicity in predominantly non-breast fed infants:		
7-mon-old German girl	Tap water, 0.4 to 5.5 mg/L	Muller-Hocker et al. 1987; Muller et al. 1998
5- and 9-mon-old siblings	Tap water, 2.2 to 3.4 mg/L	Muller-Hocker et al. 1988
10-mon-old, rural southern Ireland	Cold tap water, 3.9 mg/L; hot, 8 mg/L; copper blackened the inside of kettles	Baker et al. 1995
13-mon-old German male	Tap water, 12-29 mg/L after standing in pipes	Bent and Bohm 1995
14-mon-old Australian boy	First-draw cold water, 6.75 mg/L; hot water, 9.4 mg/L	Walker-Smith and Blomfield 1973
15-mon-old female, rural southeastern Ireland	Well-water heated and stored in copper container; hot water measured 6.3 mg/L; cold water, 2.3 mg/L.	Baker et al. 1995
Indian childhood cirrhosis	Heating and storing infant's milk in untinned copper and brass containers; simulated under experimental conditions copper at 6 mg/L in milk in copper containers, and associated dose, ≈0.9 mg/kg·d	O'Neill and Tanner 1989
Tyrolean childhood cirrhosis	Infant feeding milk prepared in untinned copper containers, in experimental simulations 10-63 mg/L	Muller et al. 1996, 1998

For example, a LOAEL associated with such a serious end point as liver failure might be divided by a factor of 3 to 10 to adjust for the severity of the end point, a factor of 10 to adjust an effect level to a level associated with no effect, and an additional factor between 2 and 10 to adjust for human heterogeneity. For conditions in which the observation is for a susceptible group a factor less than 10 might be used.

Although further research is needed to better define drinking-water doses associated with copper toxicity in those with genetic susceptibility syndromes, such as ICT and ICC, drinking-water concentrations of 3 mg/L and above are associated with hepatotoxicity in case reports (Table 6-3). However, that observation is for case reports for which cumulative intake of copper from water and other sources is not known.

The MCLG is a health goal set at concentrations at which no known or expected adverse health effects occur and the margin of safety is adequate (Chapter 1). However, it is difficult to define an adequate margin of safety for copper toxicity. The standard method for performing a safety assessment for chronic toxicity end points cannot be applied, primarily because the method does not account for certain features related to copper essentiality, tight homeostatic control, and the relatively narrow range of acceptable intake levels. Also, the dose of copper that causes liver toxicity in sensitive humans is uncertain. Nonetheless, it is important to evaluate the potential for liver toxicity from chronic exposure to copper in drinking water when considering changes to the MCLG.

CHRONIC COPPER EXPOSURE THROUGH TAP WATER

Comprehensive nationwide survey data for copper in drinking water are not available, and therefore estimates of copper intake via water cannot be estimated accurately. Clues as to the potential for copper overexposure via tap water come from federal reporting requirements. Under federal law, water systems are required to be sampled for copper in first-draw water (i.e., after water has been "motionless" for at least 6 hr) at the cold-water tap at locations in the water system vulnerable to copper contamination (EPA 1991, 1994). When the 90th percentile of samples taken exceeds 1.3 mg/L, the water purveyor is to report that percentile value to the states, which in turn are required to compile and report such values to the U.S. Environmental Protection Agency.

Figure 6-2 presents the 90th percentile copper concentrations that water purveyors reported for their systems from 1991 to 1999 (E. Ohanian, EPA, personal commun., Nov. 23, 1999). The 7,307 values reported correspond to roughly 4,500 individual water systems. With a few exceptions, water systems reporting values greater than 5 mg/L are small, serving

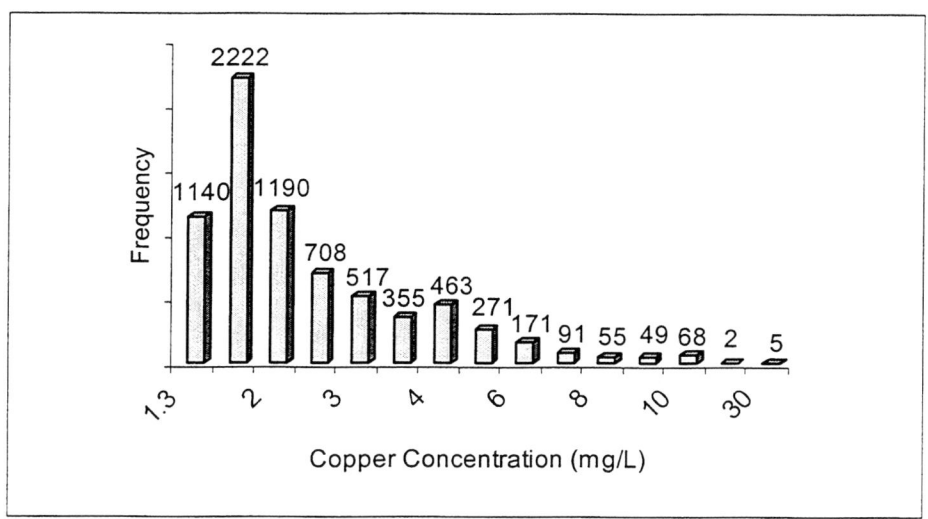

FIGURE 6-2 Ninetieth percentile copper concentrations for water systems reported to EPA under federal reporting requirements from 1991 to 1999. Source: Data from E. Ohanian, EPA, personal commun., Nov. 23, 1999.

3,300 or fewer people. The majority of those serve nonresidential consumers, such as those at recreational facilities and schools. By law, corrective action might be required for a number of those systems. Nonetheless, the reported 90th percentile concentrations for numerous systems, some of which serve small communities, are notably high, suggesting the potential for copper overexposure.

DIETARY CONTRIBUTION AND TOTAL COPPER INTAKE

Formula-Fed Infants

Formula-fed infants, particularly those receiving most sustenance from powdered formulas, are a group of particular concern. Virtually all fluid of young infants on a powdered formula diet can come from tap water, and the powder formulation is designed to provide the copper requirement.

Using information on energy requirements, copper intake via powder formula can be estimated. Infant formulas sold in the United States generally contain 75 µg of copper per 100 kcal of energy supplied by the powder. Energy expenditure and growth studies provide data on daily energy intake of infants, and data compiled (Whitehead 1995) are presented in

Figure 6-3. In infants, the energy requirement on a body-weight basis (e.g. kilocalorie per kilogram of body weight) appears to be greatest during the neonatal period and drops steadily until roughly age 6 months, when growth slows and the requirement becomes fairly constant. From those data, the average energy requirement during the first 6 months of life is roughly 100 kcal/kg per day. Copper intake associated with that energy requirement is 75 µg/kg per day (75 µg/kg per day = 100 kcal/kg-day × 75 µg/100 kcal) from the formula powder alone. Daily copper intake from infant formula made from powder and copper-contaminated tap water can be estimated. At the current MCLG of 1.3 mg/L, average daily copper intake during the first 6 months of life is estimated to be 267 µg/kg per day for the average powder-formula-fed infant.

WHO (1996), after considering copper concentrations not associated with detrimental effects in adult humans, set a value of 150 µg/kg-day as the upper limit of the safe range for mean copper intake for infants.[1] The

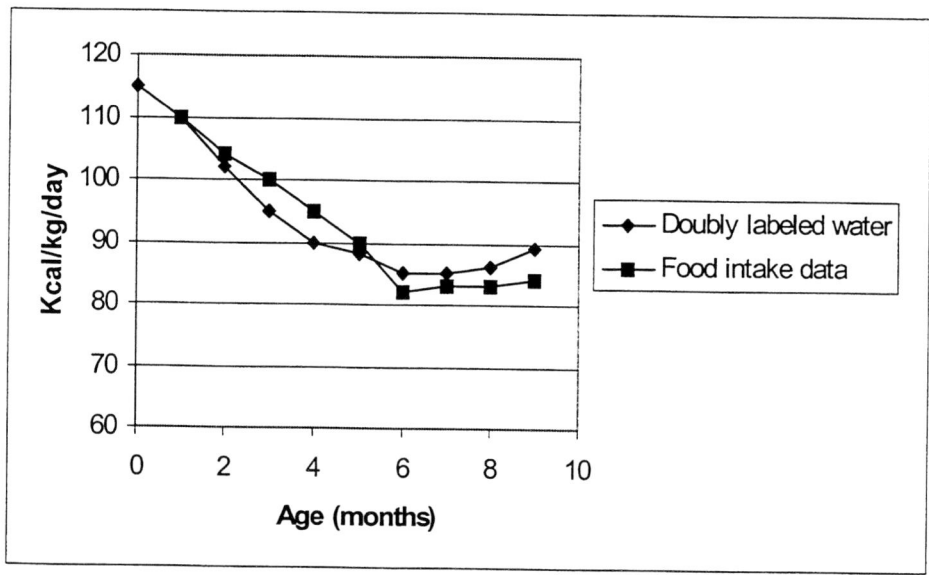

FIGURE 6-3 Energy requirements for infants predicted from doubly labeled water (following technique of Coward et al. 1979) and food-intake data. Source: Adapted from Whitehead 1995.

---

[1]The rationale and scientific basis for that value was not discussed in detail.

above value is roughly half the intake for infants on a powdered formula diet made with water containing copper at the MCLG. The copper concentration in tap water corresponding to the WHO upper limit is 0.44 mg/L for the average infant fed for the first 6 months of life on standard powder formula. Formula-fed infants consuming water contaminated at 6 mg/L would approach doses of 1 mg/kg per day, a dose associated with cases of liver toxicity in genetically sensitive infants by some researchers (Table 6-3) and approximately a factor of 10 of doses with observed effects in chronic exposure animal studies.

Consumption of water at such a high concentration corresponds to a plausible worst-case scenario for copper intake in which the day's formula is made with first-draw water. The values are calculated for the median infant. Breast-milk consumption (Hofvander et al. 1982; Dewey et al. 1991) and energy requirement (Prentice et al. 1988; Whitehead 1995) studies suggest moderate variation in intake, even when normalized to body size. Those studies suggest that the more voracious infants can consume at rates 30-50% higher than the average.

General Population

From the results of nationwide dietary surveys, copper intake from food can be estimated for different age groups and for the general population (Table 6-4). Dietary survey information can also be used to evaluate water consumption habits and variations in different age groups (Table 6-5). Assuming fixed concentrations of copper, possible intake of copper through water can be evaluated. Total copper intake through food and water can then be evaluated. Figure 6-4 illustrates total copper intake at different concentrations of copper in water. Again, at a concentration of 3 mg/L, relatively high copper intake via water can result for some segments of the population.

CONCLUSIONS

- As with any essential nutrient, low copper intake can result in pathological changes, and high intake can result in toxicity.
- GI symptoms, such as nausea and vomiting, associated with acute exposures can be seen in sensitive individuals at drinking-water concentrations of copper greater than 3 mg/L.
- Systemic toxicity is not expected with an acute exposure to copper at concentrations found in drinking water. That is true even for first-draw water (less than 30 mg/L).

TABLE 6-4 Copper Intake (µg/kg-day)[a] from Food and Supplements

Age	Sex	Percentile							
		5	10	25	50	75	90	95	99
2-6 mon	M&F	44.2	54.3	75.9	94.7	120.8	148.8	169.0	226.8
7-11 mon	M&F	30.2	36.1	56.8	76.9	101.9	128.7	51.9	200.0
1-3 yr	M&F	21.2	27.5	38.2	53.3	72.7	99.2	123.1	219.0
4-8 yr	M&F	24.7	28.5	34.9	44.3	56.7	72.9	85.8	148.5
9-13 yr	F	14.0	15.8	19.6	25.3	31.8	40.3	47.9	76.5
	M	16.4	18.8	23.1	28.9	36.4	45.3	51.4	79.1
14-18 yr	F	9.1	10.6	14.1	18.5	23.9	31.2	38.9	59.7
	M	11.6	13.2	17.1	21.7	27.6	36.3	43.4	58.3
19-30 yr	F	9.7	11.0	13.6	17.4	23.8	34.2	48.8	67.1
	M	14.3	15.6	18.4	22.3	26.9	34.1	45.1	71.2
31-50 yr	F	8.8	10.1	12.7	16.8	22.9	34.4	47.6	70.9
	M	10.3	12.2	15.5	19.5	26.1	36.3	46.5	65.2
51-70 yr	F	7.5	8.7	11.3	15.2	22.6	40.2	50.7	69.2
	M	8.2	9.8	12.7	16.8	24.4	37.1	47.0	69.3
71 + yr	F	8.0	9.3	12.2	16.2	23.0	45.1	55.9	71.4
	M	9.0	10.3	13.0	16.8	23.3	38.5	47.8	66.9
Pregnant	F	10.6	12.2	16.4	24.8	46.8	58.1	65.8	85.3
All ages	M&F	9.9	11.9	16.1	24.1	42.4	67.0	88.3	145.8

[a]Intake figures developed by Environ International Corporation and Iowa State University Department of Statistics from the NHANES III nationwide survey. Breast-feeding infants and children and eight individuals reporting greater than 150 mg of copper per day from supplements excluded from the analysis.

• Those at increased risk for chronic liver toxicity are individuals on fluid diets (e.g., formula-fed infants), or with genetically determined differences in copper metabolism, or a combination of these factors.

• In the case of sensitive populations, a substantial increase in copper intake from water increases the risk for hepatotoxicity. Therefore, in considering changes in the MCLG, the extent to which the copper in water can contribute to the overall dietary copper intake of an individual must be considered.

• Copper drinking-water concentrations of 3 mg/L and greater have been associated with cases of systemic copper toxicity, suggesting that increasing the MCLG to 3 mg/L and greater might cause an increase in liver disease in genetically susceptible populations.

TABLE 6-5 Daily Tap-Water Consumption (mL/kg)[a]

	Percentile							
Age, yr	5	10	25	50	75	90	95	99
<0.5	0	0	0	5	90	139	170	217
0.5-0.9	0	0	4	36	79	103	122	169
1-3	0	0	6	17	33	51	67	109
4-6	0	1	7	17	29	45	64	91
7-10	0	1	5	11	22	32	39	60
11-14	0	1	4	9	17	26	34	54
15-19	0	0	3	9	16	25	32	61
20-24	0	1	4	10	17	31	38	79
25-54	0	1	6	12	21	31	40	64
55-64	0	1	6	13	22	31	38	57
65+	0	0	7	15	23	31	37	52
All ages	0	1	5	12	21	33	43	77

[a]Total direct and indirect tap-water consumption estimates from the USDA 1994-1996 Continuing Survey of Food Intakes by Individuals.

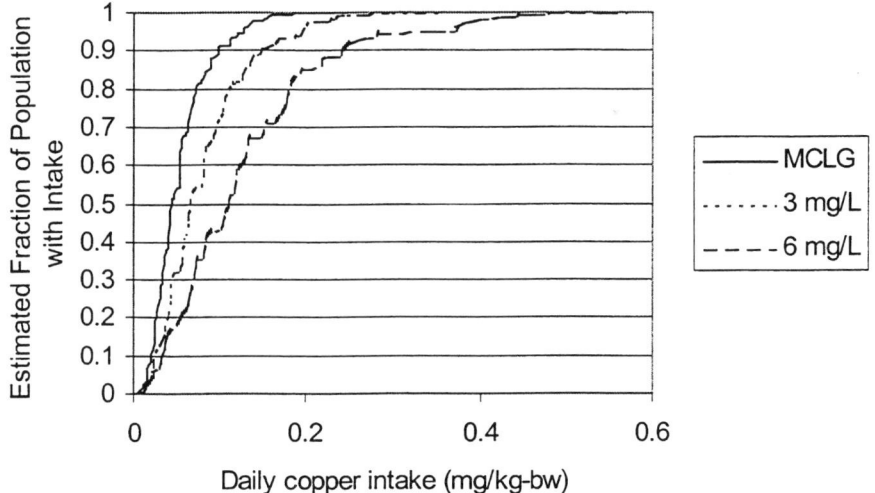

FIGURE 6-4 Total copper intake from water, diet and dietary supplements. Estimates are given for copper concentrations in drinking water at the current MCLG (1.3 mg/L; solid line), 3 mg/L (dotted line), and 6 mg/L (dashed line). Copper concentrations are assumed to be constant.

- It has been estimated that about 1% of the U.S. population have polymorphisms that increase copper retention, and might increase their risk for copper toxicosis.

RECOMMENDATIONS

- The MCLG in water should be based on the toxic effects of copper, rather than on copper deficiency.
- Issues that should be considered in establishing adjustment and uncertainty factors for acute effects are that copper is an essential micronutrient, that the GI effects are not severe or life-threatening, that the effect level is based on human studies and case reports, and that the effect level appears to be at the lower part of the dose-response curve, where the majority of the population is nonresponsive.
- Given the potential risk for liver toxicity in individuals with polymorphisms, it is recommended that the MCLG for copper should not be increased at this time.
- Additional information on total copper doses received from drinking water is necessary before the importance of systemic chronic toxicity can be fully evaluated in susceptible populations.
- Better quantification of the frequency and characterization of copper-sensitive populations should be undertaken.
- When the above information is obtained, the MCLG for copper should be re-evaluated.

REFERENCES

Baker, A., S. Gormally, R. Saxena, D. Baldwin, B. Drumm, J. Bonham, B. Portmann, A.P. Mowat. 1995. Copper-associated liver disease in childhood. J. Hepatol. 23(5):538-543.
Bent, S. and K. Bohm 1995. Copper-induced liver cirrhosis in a 13-month old boy [in German]. Gesundheitswesen 57(10):667-669.
Brewer, G.J. and V. Yuzbasiyan-Gurkan. 1992. Wilson disease. Medicine 71(3):139-164.
Cohen, J.M., L.J. Kamphake, E.K. Harris, R.L. Woodward. 1960. Taste threshold concentrations of metals in drinking water. J. Am. Water. Works Assoc. 52(5):660-670.
Coward, W.A., M.B. Sawyer, R.G. Whitehead, A.M. Prentice, and J. Evans. 1979. New method for measuring milk intakes in breast-fed babies. Lancet 2(8132):13-4.
Dewey, K.G., M.J. Heinig, L.A. Nommsen and B. Lonnerdal. 1991. Adequacy of energy intake among breast-fed infants in the DARLING study: Relationships to growth velocity, morbidity, and activity levels. J. Pediatr. 119(4):538-547.
EPA (U.S. Environmental Protection Agency). 1987. Drinking Water Criteria Document of Copper. Environmental Criteria and Assessment Office, Office of Health and Environmental Assessment, U.S. Environmental Protection Agency. Cincinnati, OH. February.
EPA (U.S. Environmental Protection Agency). 1994. Drinking water maximum contaminant level goals and national primary drinking water regulations for lead and copper. Fed. Regist. 59(125):33860-33864.
EPA (U.S. Environmental Protection Agency). 1991. Monitoring requirements for lead and copper in tap water. Fed. Regist. 56(110):26555-26557.
EPA (U.S. Environmental Protection Agency). 1998. Draft water quality criteria methodology: human health. Fed. Regist. 63(157):43756-43828.
Epstein, O., R. Spisni, S. Parbhoo, B. Woods, and T. Dormandy. 1982. The effect of oral copper loading and portasystemic shunting on the distribution of copper in the liver, brain, kidney, and cornea of the rat. Am. J. Clin. Nutr. 35(3):551-5.
Galal-Gorchev, H. and Herrman, J.L. 1996. Letter to A.C. Kolbye, Jr., editor of Regulatory and Pharmacology, on the evaluation of copper by the Joint FAO/WHO Expert Committee on Food Additives from WHO. Sept. 12, 1996.
Hofvander, Y., U. Hagman, C. Hillervik, and S. Sjolin. 1982. The amount of milk consumed by 1-3 months old breast- or bottle-fed infants. Acta Paediatr. Scand. 71(6):953-958.
IPCS (International Programme on Chemical Safety). 1994. Assessing Human Health Risk of Chemicals: Derivation of Guidance Values for

Health Based Exposure Limits. Environmental Health Criteria No. 170. Geneva, Switzerland: World Health Organization.

IPCS (International Programme on Chemical Safety). 1998. Copper. Environmental Health Criteria 200. Geneva, Switzerland: World Health Organization.

Keen, C.L. 1996. Teratogenic effects of essential trace metals: deficiencies and excesses. Pp. 977-1001 in Toxicology of Metals, L.W. Chang, L. Magos, and T. Suzuki, eds. New York: CRC Press.

Makale, M.T., and G.L. King. 1992. Surgical and pharmacological dissociation of cardiovascular and emetic responses to intragastric CuSO4. Am. J. Physiol. 263(2 Pt 2):R284-R291.

Massie, H.R. and V.R. Aiello. 1984. Excessive intake of copper: influence on longevity and cadmium accumulation in mice. Mech. Ageing. Dev. 26(2-3):195-203.

Müller, T., W. Müller, H. Feichtinger. 1998. Idiopathic copper toxicosis. Am. J. Clin. Nutr. 67(suppl):1082S-1086S.

Müller, T., H. Feichtinger, H. Berger, W. Müller. 1996. Endemic Tyrolean infantile cirrhosis: an ecogenetic disorder. Lancet 347(9005):877-880.

Müller-Höcker, J, U. Meyer, B. Wiebecke, G. Hubner. 1988. Copper storage disease of the liver and chronic dietary copper intoxication in two further German infants mimicking Indian childhood cirrhosis. Path. Res. Pract. 183(1):39-45.

Müller-Höcker, J, M. Weiss, U. Meyer, P. Schramel, B. Wiebecke, B.H. Belohradsky, G. Habner. 1987. Fatal copper storage disease of the liver in a German infant resembling Indian childhood cirrhosis. Virchows. Arch. A. Pathol. Anat. Histopathol. 411(4):379-385.

NRC (National Research Council). 1977. Drinking Water and Health. Washington, DC.: National Academy of Sciences.

NRC (National Research Council). 1989. Recommended Dietary Allowances, 10th Ed. Washington, DC.: National Academy Press.

NRC (National Research Council). 1994. Science and Judgment in Risk Assessment. Washington, D.C.: National Academy Press.

NTP (National Toxicology Program). 1993. NTP Technical Report on Toxicity Studies of Cupric Sulfate (CAS No. 7758-99-8) Administered in Drinking Water and Feed to F344/N Rats and B6C3F1 Mice. NTIS PB94-120870.

O'Donohue, J., M.A. Reid, A. Varghese, B. Portmann, R. Williams. 1993. Micronodular cirrhosis and acute liver failure due to chronic copper self-intoxication. Eur. J. Gastroenterol. Hepatol. 5:561-562.

O'Neill, N.C., and M.S. Tanner. 1989. Uptake of copper from brass vessels by bovine milk and its relevance to Indian childhood cirrhosis. J. Pediatr. Gastroenterol. Nutr. 9(2):167-172.

Pizarro, F.M., R. Olivares, P. Uauy, P. Contreras, A. Rebelo, and V. Gidi.

1999. Acute GI effects of graded levels of copper in drinking water. Environ. Health Perspect. 107(2):117-121.
Prentice, A.M., A. Lucas, L. Vasquez-Velasquez, P.S. Davies, and R.G. Whitehead. 1988. Are current dietary guidelines for young children a prescription for overfeeding? Lancet 2(8619):1066-9.
Saito, R., Y. Suehiro, H. Ariumi, K. Migita, N. Hori, T. Hashiguchi, M. Sakai, M. Saeki, Y. Takano, and H. Kamiya. 1998. Anti-emetic effects of a novel NK-1 receptor antagonist HSP-117 in ferrets. Neurosci. Lett. 254(3):169-172.
Tachibana, K. 1952. Pathological transition and functional vicissitude of liver during formation of cirrhosis by copper. Nagoya J. Med Sci. 15:108-114.
Tanner, M.S. 1998. Role of copper in Indian childhood cirrhosis. Am. J. Clin. Nutr. 67(5 Suppl.):1074S-1081S.
Walker-Smith, J. and J. Blomfield. 1973. Wilson's disease or chronic copper poisoning? Arch. Dis. Child. 48(6):476-479.
Wang, S.C. and H.I. Borison. 1951. Copper sulphate emesis: study of afferent pathways from the GI tract. Am. J. Physiol. 164:520-526.
Whitehead, R.G. 1995. For how long is exclusive breast-feeding adequate to satisfy the dietary energy needs of the average young baby? Pediatr. Res. 37(2):239-243.
WHO (World Health Organization). 1996. Trace Elements in Human Nutrition and Health. Geneva: World Health Organization.
Wyllie, J. 1957. Copper poisoning at a cocktail party. Am. J. Publ. Health 47:617.
Zucker, S. and J.L. Gollan. 1996. Wilson's disease and hepatic copper toxicity. Pp. 1405-1438 in Hepatology. A Textbook of Liver Disease, Vol. 2, 3rd Ed., D. Zakim and T.D. Boyer, eds. Philadelphia: W.B. Saunders.